Theory of Spectroscopy

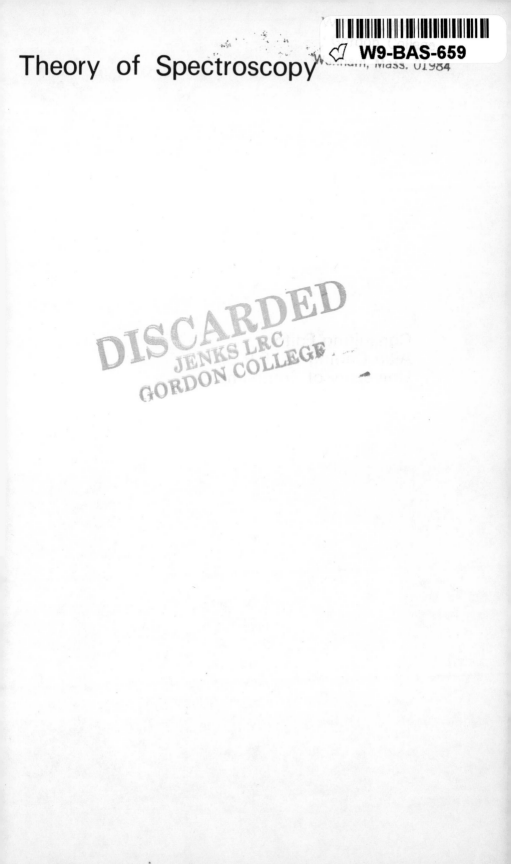

Consulting Editor
Alan Carrington, F.R.S.,
University of Southampton

Theory of

Spectroscopy

An Elementary Introduction

Oliver Howarth
University of Warwick

A HALSTED PRESS BOOK

John Wiley & Sons
New York

Published in the USA
by Halsted Press, a Division of
John Wiley & Sons, Inc., New York

First published in Great Britain by Thomas Nelson & Sons Ltd, 1973
Copyright © Oliver Howarth 1973
Illustrations © Thomas Nelson & Sons Ltd 1973

ISBN 0 470–41667–X
Library of Congress Catalog Card No. 73–35

Filmset by M. & A. Thomson Ltd., East Kilbride, Scotland
Printed in Great Britain

Contents

5 Principles underlying spectroscopic techniques

6 Quality of a spectrum

7 Nuclear magnetic and electron resonance

8 Pure rotational spectroscopy

9 Infrared and vibrational Raman spectroscopy

10 Visible and ultraviolet spectroscopy

Preface

The last fifteen years has seen spectroscopy grow from a specialist research interest into a major item in the research budget of every chemistry department. This growth has been due to the ever increasing number of practical applications of spectroscopy, particularly to the elucidation of molecular structure.

This application-led growth of interest has created a somewhat unsatisfactory situation for student teaching. Too many students are given a bare outline of the theory of spectroscopy, with most of the important results either quoted or assumed. They then have to cope with experimental details and with all the apparent exceptions to their over-simplified theoretical models. This 'classical' teaching approach has the disadvantages that were also present in the 'classical' teaching of inorganic chemistry. Whilst it permits the weaker student to apply uncritical memorizing techniques, it offers little intellectual satisfaction and insight to the enquiring student.

The present text is written in the belief that, given a rudimentary and largely qualitative understanding of quantum mechanics, the student can understand the derivation of most of the theory of spectroscopy. Just as a simple understanding of atomic and molecular orbitals and of crystal field theory welds inorganic chemistry into an exciting and unified subject, so too a simple understanding of wavefunctions, perturbation theory, and angular momentum unifies spectroscopy. These topics are presented in the text as simply as possible, consistent with the facts that they must explain.

Much of the long-term difficulty which even the best students experience with quantum mechanics is due to an inadequate understanding of its physical basis. Chapter 1 therefore deals with this basis in some detail, in a manner which will be useful as remedial reading for the older student or as a fresh approach for the near-beginner. The next three chapters then explain the existence of the simpler quantum states relevant to spectroscopy, and also the rather neglected theory of the interaction of radiation with matter. Chapters 5 and 6 attempt to unify the experimental side of spectroscopy by treating the electromagnetic spectrum as a whole, and by abstracting such topics as resolution, lock-in detection, and signal to noise ratio from the discussion of specific spectroscopic methods. This is intended to help the student to understand the reasons for selecting particular experimental methods at particular frequencies. It also facilitates discussion of newer techniques such as Fourier transform methods. Only in the final four chapters are the main spectroscopic techniques considered individually. Even here many cross-correlations are emphasized, and the theoretical background developed previously is used to clarify 'difficult' topics such as the theory of photochemistry and the u.v. spectra of solids.

Within reason, the book attempts to explain important spectroscopic results rather than just to quote them. This explaining is done either in the main text or via the problems which form an essential part of the book. In some cases, notably in perturbation theory, the text is deliberately expansive, for the neatest explanation of such matters is not always the clearest. In other cases, such as for many selection rules, new and simplified derivations are offered. For example, by realizing that the photon possesses intrinsic angular momentum the student can quickly make sense of a whole range of vibrational and rotational selection rules.

The text is not intended to be comprehensive, concentrating as it does upon insight. Its prime use should be as a teaching book, and although it is complete in itself, it is likely to be of particular value to students who have already (as is now common) met a smattering of spectroscopy and quantum mechanics in an introductory course. It could well be combined with a short practical project in one branch of spectroscopy. It should also be of interest to more experienced students of spectroscopy, who may enjoy the unconventional approach.

The units used in the text are dual; in non-trivial cases quantities are quoted in both SI and c.g.s. units. It is hoped that this will help readers to become familiar with all the units they are subsequently likely to meet. The cm^{-1} unit has been used extensively, as its future seems assured for some time yet in the absence of a realistic alternative.

So many people have helped towards this book that it is impossible to thank them all individually. I owe a particular debt to my past mentors, Dr. L. E. Sutton, Professor R. E. Richards, Professor G. K. Fraenkel, and Professor D. H. Whiffen, and to Professor A. Carrington, F.R.S., who gave considerable encouragement at the writing stage. I am also most grateful to my colleagues and students at the University of Warwick, particularly Dr D. M. Hirst, who have helped in the reading and revision of the manuscript and in related teaching courses, and to the overworked ladies, including my wife, who laboured to type it. I would be grateful to receive suggestions for future improvements and correction of the text.

O.W.H.

1 Spectroscopy and the development of quantum mechanics

Spectroscopy and quantum mechanics are inseparable. Spectroscopy is the study of transitions of a system (usually an atom or a molecule) between its states of defined energy.[1] The types of energy state with which this book is concerned, and the type of spectroscopy used in detecting transitions between states, are listed in table 1.1.

Table 1.1

Type of state	Type of spectroscopy
Subnuclear energy states	γ-ray
Inner electron shells	X-ray fluorescence
Outer electron shells	Optical and ultraviolet
Vibrational states	Infrared and Raman
Rotational states	Microwave
States of magnetic moment orientation in a magnetic field	{ Electron resonance { Nuclear magnetic resonance

The existence of these energy states is in itself a fact which cannot be adequately explained without quantum mechanics. Furthermore, spectroscopy provides a large amount of quantitative data with which the predictions of quantum mechanics are tested and the theory refined.

Conversely, the experimental facts of spectroscopy may only be understood by having a working knowledge of the underlying theory. Therefore our first task is to outline the development and principles of quantum mechanics.

1.1 Classical optical spectroscopy

The first known spectroscopic observations to distinguish individual atomic transitions were those of Josef Fraunhofer in 1814. Fraunhofer used his skill in the manufacture of optical instruments to build a spectroscope capable of resolving the lines, now named after him, which arise from absorption of solar radiation by atoms of the cooler gas which surrounds the sun. Soon the same spectroscopic analysis was applied to

---

[1] Such states of defined energy are commonly called stationary states, for reasons which are outlined in chapter 4. In this chapter there is also discussion of what one really means by transitions between 'stationary' states.

1

atomic emission from flames, and by the end of the century atomic emission spectroscopy was established as a qualitative analytical technique. The structure of the atom was not fully established until the twentieth century. However, classical physics was able to explain spectroscopic absorption, along with many optical phenomena such as scattering, refractive index, and dispersion, by assuming that the electrons in an atom had certain characteristic frequencies of oscillation, several for each electron. These frequencies are now understood to correspond to the energy differences between atomic levels.

Further progress was made once Ernest Rutherford's model of the atom, with electrons orbiting a nucleus, was accepted. By this time, owing to improvements in spectroscopic technique, a large number of lines had been observed in the spectrum of atomic hydrogen, and of other atoms. In a given region of the hydrogen spectrum the wavenumbers $\tilde{v}$ of the lines could be accurately fitted to a formula

$$\tilde{v} = v/c = R_\infty(1/n^2 - 1/m^2)$$

where c is the velocity of light, v is the frequency, n and m are integers ($m > n$), and n is fixed for a given spectral region. R_∞ is a fundamental constant, the Rydberg constant, and has the value $109\,737.3$ cm^{-1}.[2] Approximate regularities were found for other atoms. It was also noted that if two lines were observed then a third line was also sometimes observed whose wavenumber was the exact sum of the wavenumbers of the first two lines.

1.2 The Bohr atom

These facts suggested most strongly that in a transition the electrons in an atom jump between a limited number of states of defined energy, or 'energy levels', subject to certain limitations on which jumps are allowed, called 'selection rules'. However, this theory, which was first proposed by Niels Bohr in 1913,[3] involved several radical conclusions, namely:

  (i) An electron in an atom may only be bound with certain energies, which for one-electron atoms or ions are proportional to $1/n^2$ where n is integral.
 (ii) It may also only have certain angular momenta, $nh/2\pi$.
(iii) The transitions of electrons between these levels may not be understood classically.

This third conclusion was Bohr's rather absolute way of avoiding a notorious difficulty of the classical orbiting electron theory. An orbiting electron is continuously accelerating towards the nucleus, and hence according to classical electrodynamics should radiate continuously and

---

[2] Wavenumbers are commonly measured in cm^{-1}, i.e. as the reciprocal of the wavelength in cm.

[3] For a fuller discussion see B. L. van der Waerden, *Sources of Quantum Mechanics*, North Holland, Amsterdam, 1967.

thus lose energy continuously until it spirals into the nucleus. Bohr's conclusion was consistent with the fact which was then beginning to be accepted that all radiation was actually emitted discontinuously on the atomic scale.

The great strength of the Bohr theory was that it correctly predicted all the known atomic states of hydrogen with their correct energies, and thus derived Rydberg's constant from the known mass and charge of the electron, the velocity of light, and Planck's constant.[4] Its weakness was partly that it did not really explain the quantization it assumed, except perhaps by analogy with Planck's theory, and partly that most of its quantitative predictions of angular momentum were incorrect. In particular, it ruled out the possibility of states with zero angular momentum, which we now know as s-states. Thus it was unable adequately to explain the effect of a static magnetic field on atomic spectra (the Zeeman effect). However, its success in predicting energy levels justified its disregard for classical physics and thus opened the way for the even more radical theory of quantum mechanics.

1.3   Vibrational quanta

The original suggestion of quanta of radiation came from Max Planck's investigations into black body radiation,[5] and was announced in 1900. A black body is a body which both absorbs and emits the maximum possible amount of radiation at all frequencies. No real surface is ideally 'black' for all wavelengths, for precisely the reasons that make spectroscopy interesting. But the very small exit of an otherwise closed cavity is very nearly 'black' at all wavelengths, because once radiation gets inside the cavity it must undergo very many potentially absorptive reflections before escaping again. An equilibrium argument shows that all black bodies at any one temperature must emit the same amount of radiation per unit area at any one frequency, and hence that a plot of energy density versus frequency of radiation from a black body is independent of the particular body and thus is of fundamental physical significance. Such a plot is shown in figure 1.1. Before Planck's work there were two main theories about black body radiation. The first was the strict classical theory, which considered the body as being made up of harmonic oscillators of frequency v. The only limitation on v was that the resulting radiation (which, as explained in chapter 3, would necessarily be of frequency v) had to be a standing wave within the black body cavity. The conclusion was that

$$u = \frac{8\pi v^2 kT}{c^3} \tag{1.1}$$

where T is the absolute temperature, u is the energy density per unit

[4] See section 1.3.
[5] See M. Jammer, *The Conceptual Development of Quantum Mechanics*, McGraw-Hill, New York, 1966, chapter 1.

frequency range, and k is the gas constant divided by Avogadro's number. This classical theory correctly predicted the low frequency part of the curve in figure 1.1, but unfortunately also predicted that this rising portion of the curve would continue to rise indefinitely, with increasing frequency. This conflict between experiment and classical theory became known as the 'ultraviolet catastrophe'.

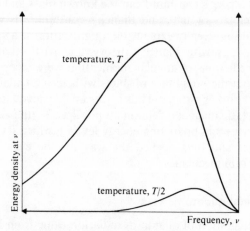

Fig. **1.1** Typical experimental plot of radiation energy from nearly black body against ν

As well as knowing the classical radiation law, Planck also knew the Wien law, which was

$$u = \frac{8\pi h\upsilon^3}{c^3}e^{-h\nu/kT} \tag{1.2}$$

This fits the experimental curve very approximately, the fit improving as υ increases. However, it was justified by rather doubtful arguments including an analogy with the molecular speed distribution in gases.

Planck took the bold step of combining equations 1.1 and 1.2 into one formula

$$u = \frac{8\pi\upsilon^2}{c^3}\cdot\frac{h\upsilon}{e^{h\nu/kT}-1} \tag{1.3}$$

which may readily be seen to reduce to 1.1 and 1.2 in the low and high frequency regions respectively, and which also fits experiment exactly. But he could only justify his formula theoretically by assuming that radiation existed in discrete units of energy, called 'quanta' or 'photons', such that their energy ε was related to their frequency υ by the relationship

$$\varepsilon = h\upsilon \tag{1.4}$$

where h, Planck's constant, has a magnitude of 6.625×10^{-34} J s ($= 6.625 \times 10^{-27}$ erg s). He simultaneously concluded that the oscillators could only accept or lose energy as these quanta, and hence that an oscillator of frequency v could only have energies 0, hv, $2hv$, etc. Later, Peter Debye pointed out that although these steps of energy were correct, their zero was not actually defined by the theory. We now know that the 'zero' is in fact $\frac{1}{2}hv$ and hence that the levels have energy $\frac{1}{2}hv$, $\frac{3}{2}hv$, $\frac{5}{2}hv$, etc. This proof of the existence of vibrational energy levels encouraged Bohr to develop his quantum theory of electronic energy levels, and also led Niels Bjerrum to a correct interpretation of the infrared absorption spectra which he first described in 1912.

1.4   Boltzmann's law

So far we have not explained why Planck's hypothesis succeeded in rectifying the 'ultraviolet catastrophe' objection to the classical theory of black body radiation. In order to understand this we must study Boltzmann's law as it applies to systems with discrete energy levels.[6] Boltzmann's law states[7] that the ratio of the number n_1 of particles with energy ε_1 to the number n_2 of energy ε_2 in any non-degenerate energy level in any degree of freedom is given by the relation

$$n_1/n_2 = e^{-(\varepsilon_1 - \varepsilon_2)/kT} \tag{1.5}$$

when normal thermal equilibrium has been established. From this it follows that if $(\varepsilon_1 - \varepsilon_2) \gg kT$ then n_1/n_2 will be vanishingly small. Therefore energy levels which are much more than kT above the lowest available level will not be significantly filled under conditions of thermal equilibrium. At room temperature kT/h is about 6×10^{12} Hz, corresponding to very low frequency vibrations.

Planck's hypothesis avoids the ultraviolet catastrophe because those oscillators whose frequency v is well above kT/h will be most unlikely to exist, according to Boltzmann's law, in any energy state other than their lowest. For equation 1.4 requires that their first excited state be an energy hv above their ground state. As the high frequency oscillators will not be able to radiate from their ground state they will not contribute to the black body radiation spectrum.

Boltzmann's law may be quantitatively verified by the study of specific heats. At 0 K all molecules must be in the ground state in all their degrees of freedom. As they warm up, they are progressively able to fill higher and higher energy levels. Each time one molecule changes from a lower

[6] It is common spectroscopic parlance to describe a rotational, vibrational, electronic, or other quantum state as an 'energy level' and to describe molecules in a particular state as 'filling' that state. 'Degenerate energy level' is also commonly used to describe a degenerate group of quantum states, i.e. a group having the same energy.

[7] Boltzmann's law is stated here in a form which is appropriate for a system with quantized, non-degenerate energy levels filled with distinguishable particles. It can be derived much more generally, and is of primary importance in statistical thermodynamics. See for example G. Sonntag and G. Van Wylen, *Fundamentals of Statistical Thermodynamics*, Wiley, New York, 1966, for a relatively straightforward derivation. It is also only valid for systems with a large number of molecules, as its origin is statistical.

to a higher level it absorbs just precisely the energy corresponding to the separation of those levels. The number of molecules filling any particular level is fixed, relative to the total number of molecules in the system, by Boltzmann's law, and hence the total internal energy E of the system may be calculated as a function of T provided that the energy levels are known. The specific heat (at constant volume) $= (\partial E/\partial T)_V$, and hence this may also be calculated and shown to fit with experiment, thus supporting Planck's quantum hypothesis. Let us consider two qualitative examples.[8]

(i) Electronic states of an atomic gas

In most cases the electrons in an atom make a negligible contribution to its specific heat at normal temperatures. This is because at these temperatures nearly all excited electronic states are separated from the ground state by an energy much greater than kT. Therefore, at equilibrium, a negligible number of excited states are thermally populated, so that the electrons are unable to absorb thermal energy and contribute to the specific heat.

One further consequence of this negligible population of upper states is that in the optical absorption spectra of cold substances one only observes transitions from the ground state. This greatly simplifies the interpretation of such spectra.

(ii) Vibrational energy states of a molecular gas

These states are normally separated at room temperature by an energy of a few times kT. Thus they begin to be significantly populated as the gas is heated above room temperature. The energy absorbed in populating them determines the vibrational contribution to the specific heat of the gas. The marked temperature dependence of the specific heat of molecular gases was never adequately explained by classical theory, but it is easily explained by quantum theory. In fact, after Bjerrum had first measured the infrared absorption spectrum of CO_2 he immediately went on to use his energy-level data to calculate the specific heat of the vapour.

The infrared absorption spectrum of a cold substance usually only shows transitions upwards from the ground state. However at higher temperatures new absorptions (hot bands) occur due to transitions from excited states with significant thermal population. This directly illustrates Boltzmann's law.

As will be seen, gas molecules are also confined to rotational and translational energy states. However, these are usually so close together in energy that they approximate to a classical energy continuum. One exception to this is hydrogen, whose rotational states are separated by exceptionally large energies. The specific heat of hydrogen is for this reason anomalous in comparison with other gases.

[8] See also problems 1.4 and 1.13.

1.5   Other evidence for the existence of photons

Planck's quantum hypothesis was finally accepted when studies of the photoelectric effect also led to the same conclusion. When light of a certain frequency, v, is directed onto an alkali metal then electrons are found to be emitted, instantly, provided hv is greater than the work function, or bulk ionization potential, ϕ, of the metal. If $hv < \phi$ then no emission at all is observed. Furthermore, the speed of the ejected electron is proportional to $hv - \phi$ and does not depend on the radiation intensity. Increasing the radiation intensity merely increases the number of electrons ejected. These results are fully consistent with the idea that the radiation reaches the metal as quanta of energy hv. But they are quite inconsistent with the classical theory in which the incident radiation, whatever its frequency, removed some of the metal's electrons after a suitable time lag by some undefined process of energy concentration.

Similar evidence came from studies of the Compton effect. According to classical theory, the effect of incident X-rays on free or weakly bound electrons should be to subject them to a steady radiation pressure and hence acceleration away from the direction of incidence. The electrons are actually ejected discontinuously at angles which are consistent with the radiation arriving as photons. The frequency shift of the scattered radiation in all directions is also consistent with this collision model and not with shifts predicted by the classical theory.[9]

1.6   Diffraction evidence for the wave nature of particles

The observation that electrons and atoms are restricted in their possible energies is now understood to be a consequence of their wave nature. The wave nature of particles had, however, to be demonstrated more directly before it could be used as a basis for quantum mechanical theory. Such a demonstration was provided when electron diffraction was observed by C. Davisson and L. H. Germer and by G. P. Thomson and A. Reid. The diffraction is closely analogous to the diffraction of light from a grating, and an electron with momentum p behaves in a diffraction experiment as if it were radiation of wavelength $\lambda = h/p$. The relationship

$$\lambda = h/p \tag{1.6}$$

describes a quite general property of particles, and is known as the de Broglie relationship because it was originally predicted theoretically by Louis de Broglie as a part of his attempt to explain Bohr's postulates. Electron diffraction and neutron diffraction are now established techniques for structure determination.

In order to gain further insight into the wave nature of matter, let us consider an idealized Davisson and Germer experiment. In figure 1.2

---

[9] The analysis of the Compton effect is not easy, but a useful account is given in M. Karplus and R. Porter, *Atoms and Molecules*, Benjamin, New York, 1970, pp. 55–57.

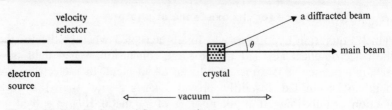

Fig. **1.2** Idealized Davisson and Germer experiment

most of the electron beam goes straight through the crystal, but some electrons are scattered at angles such as θ which depend only on the momentum of the electrons and on the specific structure of the crystal. By measuring θ for a crystal of known structure, it is possible to measure the momenta of the electrons in the beam.

Let us now consider the results of two further experiments using this electron diffraction apparatus. The results of these experiments are of primary importance in quantum mechanics.

The first experiment is to insert a second, identical, crystal into part of the diffracted beam from the first crystal. The result of this turns out to be exactly as in the original experiment; part of the original diffracted beam is diffracted again through a further angle θ. This shows that the electron momentum is unaltered in magnitude. Therefore there are at least some electron properties which can be repeatedly observed without their being altered by the process of observation.

The second experiment complements the first by showing that despite this, not all electron properties are unchanged by the process of observation. The experiment attempts to measure the position, x, of the electron at a given moment in time and subsequently to measure the momentum of the electron. A rotating metal disc with a small slit in it is inserted in the original beam. The disc is earthed, so that for most of the time the electrons simply strike it and are absorbed. However, at certain known moments in time the slit in the disc permits the electron beam to pass through and strike the crystal. Thus a length of beam, Δx, is permitted to pass, and the electron position is known with this accuracy. Such a chopped wave might look as in figure 1.3.

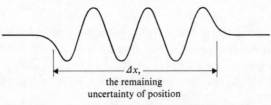

Fig. **1.3** Chopped wave containing electrons within a range Δx

A single wavelength is only precisely defined when it describes the peak to peak distance of an infinitely extended pure sinusoidal wave. Any other repeating wave pattern may, however, be analysed as the sum of many superimposed pure sinusoidal waves.[10] For example, the wave pulse in figure 1.3 may be imagined as containing the original unchopped wavelength plus many nearby wavelengths which interfere constructively at the centre of the pulse but destructively before and after the pulse. The distribution of frequencies in the pulse would be approximately as shown in figure 1.4, and the longer the pulse, the sharper would be the main peak. One can express this by saying that an unavoidable uncertainty Δv has been introduced into the measurement of frequency, where Δv is defined as the root mean square deviation of the frequency, and is inversely proportional to Δx.[11] The more closely we attempt to measure x, the more error we introduce into our measurement of v. Thus the attempt to fix the approximate position of the electrons at a given moment

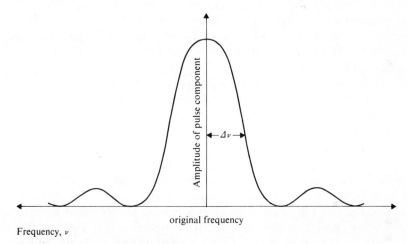

original frequency

Frequency, v

Fig. **1.4** Frequency spread in pulsed wave

and subsequently measure their precise momentum by Davisson and Germer's experiment fails. This is, of course, contrary to the assumption of classical mechanics that all physical properties are in principle precisely measurable. It shows the need for a new form of mechanics which does not make this assumption.

1.7   Evidence for the quantization of angular momentum

The angular momentum of a spinning body remains constant provided that no couple acts on the body. Thus angular momentum is a fixed property of bodies such as atoms, nuclei, and electrons in so far as these

[10] See G. Stephenson, *Mathematical Methods for Science Students*, Longmans Green, London, 1961, ch. 15. The underlying theory is known as Fourier analysis.

[11] This argument is also fully valid for a classical wave, and leads to the relation $\Delta v \Delta t \geqslant 1/4\pi$. Thus a musical note emitted for $\frac{1}{10}$ second is only defined to within about one hertz either way. For this reason even an ideal instrument cannot produce low bass 'runs' without unavoidable blurring of the sound.

are isolated. Furthermore, the amount of angular momentum is directly related by both classical and quantum electrodynamics to the magnetic moment that results from the rotating charges. One component of the magnetic moment can be measured, for example by observing the amount of deflection of the particle in an inhomogeneous magnetic field.

Stern and Gerlach carried out just this experiment on a beam of silver atoms. They expected, on classical grounds, that the beam would widen into a band with maximum density at the centre, corresponding to an almost random thermal equilibrium distribution of the magnetic moment orientations of the atoms (see section 1.8 (iii)). Instead, they found that the beam split into two, one beam being deflected about half the maximum possible amount towards the direction of higher field, and the other being deflected exactly the same amount in the other direction. When the experiment was repeated on other atoms the results were very similar, except that with other atoms the beam split into more equally spaced beams or did not split at all. It was never deflected by the full amount permitted by classical theory.

These results showed that the component of angular momentum of the atoms along the direction of the field was quantized. Just as electrons may only have certain energies in atoms, so too may atoms only have certain orientations of their angular momentum relative to a magnetic field. This conclusion fitted neatly with what was then known about the Zeeman effect in optical spectra, i.e. the line splittings observed in an optical spectrum when the sample is put into a static magnetic field. A detailed study of these splittings (see section 10.5) had led not only to the conclusion that the atomic angular momentum was confined to certain directions relative to the magnetic field, but also that electrons have intrinsic angular momentum, or spin, in addition to any angular momentum they may acquire as a result of their orbital motion. This additional conclusion was confirmed by Stern and Gerlach's experiment, because the electrons in the silver atom do not have any net orbital angular momentum and therefore their observed magnetic moment can only be due to spin.

Later experiments confirmed the existence of nuclear spin, and both experiment and theory affirm that all charged bodies with angular momentum show the above 'space quantization' in a magnetic field. The reason for this will be considered further in the next section.

1.8 The uncertainty principle

It was shown in section 1.6 that a diffraction experiment could not be used to measure both the precise position and the precise momentum of one or more electrons. An attempt to define the position along one coordinate to within a range Δx led to an uncertainty in momentum along that coordinate, Δp_x. Furthermore Δp_x was inversely proportional to Δx. A more general analysis shows that $\Delta x \Delta p_x \geqslant h/4\pi$ (1.7), where

Δx, Δp_x are understood to be the root mean square deviations of x and p_x.

In 1926, Werner Heisenberg proposed that the above relation was invariably true, for all particles. It is known as his uncertainty principle. Let us look at another idealized experiment which also demonstrates the principle. Suppose that we attempt to measure the position and momentum of an atom by striking it with a photon at two known moments. The photon will be scattered and by extrapolating backwards to the source of the scattering the two positions and hence the momentum may be calculated. Unfortunately, however, there is a limitation on the accuracy of this extrapolation which approximately equals the wavelength of the photon. (This is a general consequence of classical optics. It imposes a fundamental limitation on the resolving power of optical microscopes.) One can, of course, reduce the uncertainty in position (Δx) by reducing the wavelength of the photon. But a photon of wavelength λ imparts a momentum h/λ of indeterminate direction to the particle with which it collides. Thus the particle acquires an uncertainty in momentum $\Delta p \simeq h/\lambda$ each time it is located within a range $\Delta x \simeq \lambda$. Clearly $\Delta x \Delta p \simeq h$, which is consistent with the Heisenberg principle.

At first many eminent physicists, including Einstein, disputed the general truth of Heisenberg's principle. But no practical, or even idealized experiment has ever been devised which has succeeded in disproving it, provided that full account is taken of the quantum nature of radiation and also that of the measuring devices proposed. Quantum physics is now based on the principle that the overall properties of matter are not independent of how they are observed. As a consequence, at least some of these properties cannot be accurately measured without simultaneously changing other properties of the system so that their values are no longer known. Thus the apparent paradox in the statement 'an electron behaves both as a wave and as a particle' is resolved by a more precise use of language. If an electron behaves like a classical wave, it has a precise momentum and a completely unknown position. If it behaves like a classical particle it has a precisely known position but a completely unknown momentum. The precise measurement of the one property changes the other property so that any previous measurements of it are rendered useless. The 'nature' of the electron at any moment depends on the measurement to which it has been subjected, and thus an electron may have the incompatible properties of a wave and of a particle just because it may only have these one at a time.

The uncertainty principle may be directly applied or expanded in order to explain certain spectroscopic observations.

(i) Application of the momentum–position relation

Classical theory could not explain why an orbiting electron did not spiral down into the nucleus and become captured. However, if it did so spiral its position and velocity would become as well defined as those of the much heavier nucleus, and hence its momentum would be far more precisely defined than the uncertainty principle permits.

A more detailed analysis (see problem 1.5) shows that the uncertainty principle approximately defines a minimum radius for the electron orbit, and that this radius is close to the experimental average radius. A similar one-dimensional analysis of an atom confined by a chemical bond shows that it may never be at rest in one position. There is always some vibrational energy, called 'zero point energy', even at the absolute zero of temperature.

(ii)   Extension of the principle to angular momentum–angle

Let us suppose that a particle of mass m and momentum p is moving in a circular path at distance r from an axis O, and that its position at a given moment is described by the angle θ between the particle-axis radius and some reference radius. This is illustrated in figure 1.5. Now the

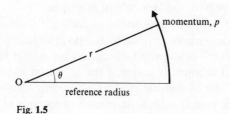

Fig. **1.5**

position of the particle, measured as the distance along the path from where it is cut by the reference radius, is $r\theta$, and the angular momentum of the particle is equal to its moment of inertia about O, mr^2, multiplied by its angular velocity p/mr, i.e. equals pr. Now the uncertainty principle requires that $\Delta p\,\Delta(r\theta) \geqslant h/4\pi$, and because r is constant we may rewrite this as $\Delta(pr)\Delta\theta \geqslant h/4\pi$. In other words, the same uncertainty relationship exists between angular momentum and angle about any one axis, as exists between position and momentum along any one coordinate.

This leads to an understanding of molecular rotation. It tells us that we can only measure rotational energies (and hence angular momenta) precisely by forgoing any information about the orientation of the molecule viewed down its axis of rotation. Any attempt to measure such an orientation would require a photon of wavelength shorter than or approximately equal to the molecular size, and such a photon would impart so much random angular and linear momentum to the molecule as to leave its rotational energy virtually undefined. This sort of argument further underlines the fact that the language of quantum mechanics is just as necessary to describe atoms and nuclei correctly as it is to describe electrons correctly.

(iii)   Extension of the principle to separate components of angular momentum

Let us suppose in the above rather simplified example that the angular position of the particle is completely unknown, and that the angular momentum is precisely defined. In order for this model to exist we had to specify a constant radius, and thus the momentum along the radial direction must have been entirely undefined. Therefore the angular momentum about any other axis would also be completely undefined, as at least one of its components would be unknown. This illustrates a general rule which states that no two components of the angular momentum of a body may be precisely and simultaneously measured. We may, however, know the magnitude of the total angular momentum of the body: this is often a fundamental property.

Another way of expressing this is to say that although the total angular momentum of a body may be known, the exact direction of the rotation axis may not be measured. (If it could be measured, we would at once know all three components of the angular momentum about any coordinate axes that we chose and hence the earlier principle would be violated.) The most that we may measure precisely is the total angular momentum and one component of it.

The physical consequences of this rule may be illustrated by a further analysis of Stern and Gerlach's experiment. If classical physics applied, the magnetic dipoles of the atoms would at normal temperatures point almost randomly in relation to the direction of the applied field. The necessary potential energy would be readily supplied by exchange with thermal energy. Thus some atoms would have their dipoles pointing almost exactly along or against the field direction, and as the field is also inhomogeneous they would be deflected to form the top and bottom of the split beam. Their deflection would thus measure their total angular momentum and show that it pointed along or against the field direction. This clearly violates the above version of the uncertainty principle, and hence it is not surprising to find that these extreme orientations are not observed.

Stern and Gerlach's experiment measures one component of the angular momentum of an atom. One might consider the possibility of repeating the experiment on a split part of the beam, putting the field direction at right angles to that in the first experiment. This possibility is, however, ruled out by the fact of precession. According to both classical and quantum mechanics a magnetic dipole which is not pointing exactly along or against the direction of a magnetic field behaves in the same way as a spinning top tilted from the vertical. It precesses about the field direction at a rate determined by the size of the field and the rate of spin. This serves to ensure that its angular momentum component in any direction normal to the field fluctuates, and hence has no precise, fixed value. Thus a repeat of Stern and Gerlach's experiment with a new field direction invalidates the information gained from the earlier experiment. The precession rate, measured as a frequency, is known as the Larmor precession frequency. The existence of precession also explains why even a classical particle with a magnetic moment does not immediately align when put in a magnetic field. It precesses instead.

(iv)   Energy and time

The position–momentum uncertainty relation was illustrated in section 1.6 by considering a beam of electrons emitted for a very short time. If the wave velocity in this experiment is c, and the pulse length Δx, this means that electrons are permitted to escape for a time interval $\Delta t = \Delta x/c$, and thus that the time of their escape is only known within this limit. Their momentum has an uncertainty Δp. Now $v = c/\lambda$ and $p = h/\lambda$. Therefore there is an uncertainty in frequency of $c\Delta p/h = \Delta v$ and a consequent uncertainty in energy of $c\Delta p = \Delta \varepsilon$. But we know that $\Delta x \, \Delta p \geqslant h/4\pi$, and hence in this case $\Delta \varepsilon \, \Delta t \geqslant h/4\pi$. This new uncertainty relationship is quite generally valid, although the above derivation is not, and will be improved in chapter 4. It tells us that the energy of a particle becomes more and more precisely defined as the moment at which it has that energy becomes less and less defined. This may be used to explain pressure broadening in rotational spectroscopy (section 6.2 (iv)) and the broadening of n.m.r. resonances due to spin–lattice relaxation (section 7.4). It is also of use in some kinetic studies.

1.9   Wavefunctions and operators[12]

The development of quantum mechanics from experiments such as have been outlined relied on the work of many physicists including Heisenberg, E. Schrödinger, P. A. M. Dirac, and W. Pauli. To justify their hypotheses they drew on many principles of classical physics and of mathematics such as Hamiltonian mechanics, matrix algebra and the theory of vibrations. The following account attempts to show only the underlying logic of quantum mechanics and not its detailed justification, which in any case ultimately rests on its successful application.

This chapter has already outlined the discovery of several important new facts about waves and particles. For example,

a   The properties of a particle at any one instant are never all known precisely. This conflicts with the assumptions of classical mechanics and means that in order to describe nature correctly we must establish a new form of mechanics.

b   In general, any measurement will affect the properties of the particle.

c   However, certain measurements do not affect its properties under

[12] The approach given here is along the lines of the much more detailed argument in A. Messiah *Quantum Mechanics* (vol. 1), Wiley, New York, 1958. A rather simpler account is given in P. W. Atkins, *Molecular Quantum Mechanics*, OUP, London, 1970, particularly in ch. 4.

certain conditions. These conditions are that the measured property
is exactly defined, and that at least one other property is totally in-
definite. For example, the momentum of a particle may be measured
precisely as many times as is desired provided that its position is
totally unknown. On a crude, semi-classical picture one might argue
that all measurements affect the particle, but that in some cases the
effect is on an already indeterminate property and is hence not
observable.

Fact **a** is dealt with by proposing that the properties of the particle
(or wave) are fully described by a wavefunction, conventionally written
as ψ. Consider figure 1.4. The frequency associated with the electrons
was not precise, but the distribution of frequencies was a definite curve.
Similarly a spread of possible positions could be represented by a definite
curve, or by the corresponding algebraic function. For a single particle,
ψ is normally written as a function either of cartesian coordinates and
time, $\psi(x, y, z, t)$ or of spherical coordinates and time, $\psi(r, \theta, \phi, t)$ (see
figure 1.6). It is defined as that function of coordinates which when
multiplied by its complex conjugate and evaluated at specified values of
its coordinates gives the probability of the particle's having those coordinates. Let us consider a simple case where ψ is a function of x alone,
such as in figure 1.7. Here $\psi^*(0)\psi(0)$ is the probability of the particle
being at x_0.[13] More precisely,

$$\psi^*(0)\psi(0)\,dx \bigg/ \int_{-\infty}^{\infty} \psi^*\psi\,dx \qquad (1.7)$$

is the fractional probability of the particle being found within a very small
range dx spanning x_0. If a wavefunction is such that $\int \psi^*\psi\,d\tau = 1$, where

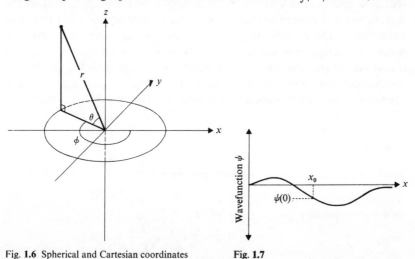

Fig. **1.6** Spherical and Cartesian coordinates   Fig. **1.7**

[13] $\psi^*(0)$ is the complex conjugate of $\psi(0)$. The reason for introducing complex wavefunctions is outlined at the beginning of chapter 4.

$d\tau$ is shorthand for integration over all possible values of all space coordinates, then it is said to be normalized. Clearly, therefore, if any wavefunction ψ is divided by the square root of $\int\psi^*\psi\,d\tau$ it will become normalized. So that ψ should describe real properties it must, like any real distribution, be a continuous, single-valued function which also has a finite integral over all space coordinates as implied above. Normally this integration condition means that ψ must be finite everywhere and must tend to zero as its coordinates tend to infinity. The extreme cases of 'pure wave' and 'pure particle'·are however exceptions to this. The reason for multiplying ψ by its complex conjugate is that even though ψ may be complex, $\psi^*\psi$ is always real.

If a wavefunction is fully known, including its time dependence, then one may calculate not only the present properties of the particle, but also its future properties under the influence of any known potentials or measurements. This is of vital importance in the theory of transition probabilities in spectroscopy.

Facts **b** and **c** are allowed for in quantum mechanics by the use of algebraic operators. An algebraic operator such as $\times 3$, $\div x$ or $\partial/\partial t$ may be applied to an algebraic function such as ψ to give a new function such as 3ψ, ψ/x, or $\partial\psi/\partial t$. Occasionally a particular operator will act on a particular function to give the same function multiplied by a constant. An example is d/dx, which operates on functions ae^{mx} (a and m constant) to give ame^{mx}. If ae^{mx} were a normalized wavefunction, ame^{mx} would be an identical wavefunction when renormalized. Such functions are known as eigenfunctions of the operator in question, and the constant m is known as an eigenvalue.

Every possible measurement or observation in the particle is represented in quantum mechanics by an algebraic operator, $A_{(op)}$. Just as the wavefunction is chosen to give an accurate description of what one may know about the particle, so too the operators are chosen to give an accurate description of the known results of the corresponding measurements on the particle. They are therefore consistent with the de Broglie relationship and with the uncertainty principle. A selection of actual operators which do this are given in table 1.2. In this table, $\hbar \equiv h/2\pi$.

Table 1.2

Measurement	*Corresponding operator*
position along x-axis	$\times x$
linear momentum along x-axis	$-i\hbar\,\partial/\partial x$
energy	$H \equiv -\dfrac{\hbar^2}{2m}\left(\dfrac{\partial^2}{\partial x^2}+\dfrac{\partial^2}{\partial y^2}+\dfrac{\partial^2}{\partial z^2}\right)+V$
z-component of angular momentum	$-i\hbar\,\partial/\partial\phi$
square of total angular momentum	$\dfrac{-\hbar^2}{\sin\theta}\dfrac{\partial}{\partial\theta}\left(\sin\theta\,\dfrac{\partial}{\partial\theta}\right)-\dfrac{\hbar^2}{\sin^2\theta}\dfrac{\partial^2}{\partial\phi^2}$

In fact all possible measurement operators may be expressed as some combination of the first two operators. Thus the kinetic energy of a classical body equals $(p_x^2 + p_y^2 + p_z^2)/2m$, and if V is its potential energy then its total energy E (strictly speaking, in a potential field such that V is independent of p_x, p_y, p_z) is given by $E = [(p_x^2 + p_y^2 + p_z^2)/2m] + V$. Now if p_x is replaced by $-i\hbar\partial/\partial x$ and thus p_x^2 by $-\hbar^2\partial^2/\partial x^2$, and p_y and p_z are similarly replaced, we obtain the energy or 'Hamiltonian' operator above. All these operators have the additional property of being Hermitian. This means that

$$\int \psi_1^* A_{(op)} \psi_2 \, d\tau = \int \psi_2 A_{(op)}^* \psi_1^* \, d\tau = \left[\int \psi_2^* A_{(op)} \psi_1 \, d\tau \right]^*,$$

and ensures that all possible observations lead to real results even when ψ is complex.

If a measurement is done on a particle whose wavefunction is ψ, it will in general affect that particle so that it subsequently has a new wavefunction ψ_{new}. The operator $A_{(op)}$ corresponding to such a measurement is in fact chosen so that

$$\psi_{new} = A_{(op)}\psi \tag{1.8}$$

As ψ will not usually be an eigenfunction of $A_{(op)}$, ψ_{new} will not usually equal ψ, even when normalized, and thus the model allows for the disturbance produced by the measurement. The special case where ψ is an eigenfunction of $A_{(op)}$ corresponds to an exact and hence repeatable measurement (condition **c**).

The uncertainty principle is implied by the fact that $\times x$ and $-i\hbar\partial/\partial x$ can never have the same eigenfunctions and can thus never be measured simultaneously and precisely. This may be made more clear, at the expense of rigour, by considering the successive measurement of position and momentum. If the initial wavefunction is ψ, then after the measurement of position (which may or may not lead to an exact result) it is $x\psi$, and after the momentum measurement it is

$$\left[-i\hbar\partial \frac{(x\psi)}{\partial x} = -i\hbar \frac{\partial \psi}{\partial x} - i\hbar\psi \right]$$

However, if the order of the operations were reversed the end result would be a wavefunction $-i\hbar x \, \partial\psi/\partial x$. The two results differ by the term $-i\hbar\psi$, and hence the state of the system after the measurements depends upon the order in which they are carried out. It would therefore clearly be a logical impossibility to carry them out simultaneously. The operators are said not to commute. Conversely it may be shown that if any operators do commute then they hold all their eigenfunctions in common. It may also be shown that the operators in table 1.2 are fully consistent with the quantitative statements of the uncertainty principle.

1.10   Average results of measurements

The operators not only describe the effect of a measurement on the particle but also its average result. Before this is explained a point of clarification is needed. The operators so far quoted correspond to idealized measurements, i.e. measurements which find the maximum available information about the relevant property of the original system and which have the minimum, unavoidable effect on it. Now an actual measurement might affect the system much more, and at the same time appear to give more precise information. For example, we might direct very short wavelength X-rays at an atom, eject an electron as in the Compton effect, and by analysing the diffracted X-rays deduce the apparent position of the electron within the atom very precisely—more precisely than quantum mechanics would appear to allow. The word apparent is used deliberately. For what we are really measuring is the position of the electron not in the isolated atom but in the combined system of atom plus moving photon. We cannot avoid this fact by arguing about events 'just before' the collision being independent of it because, if we specify the wavelength (i.e. energy of the photon) in order to interpret the diffraction results, we are no longer able to define the moment of collision with sufficient precision to deduce the electron position 'just before' that moment. The position of the electron in the *isolated atom* is thus correctly described by a spread-out wavefunction even though a rather violent experiment could change the system and thus localize it.

Let us return to figure 1.7, and ask the question 'Where is the centre of gravity of the electron cloud?' This centre of gravity might equally well be described as the mean position $\bar{x}$ of the electrons along that axis, and it is given in classical mechanics by the formula

$$\bar{x} = \frac{\int_{-\infty}^{\infty} x\psi\psi^* \, dx}{\int_{-\infty}^{\infty} \psi\psi^* \, dx} \tag{1.9}$$

This formula is derived by finding the total moment of the cloud about the centre of coordinates and dividing by its mass. To find the average value of a property in quantum mechanics one simply substitutes the appropriate operator as in table 1.2 in place of x. If this operator is $A_{(op)}$, and the average value of the observable A is $\bar{A}$, then in general

$$\bar{A} = \frac{\int A_{(op)}\psi\psi^* \, d\tau}{\int \psi\psi^* \, d\tau} \tag{1.10}$$

In the simple case of x and $\bar{x}$ this reduces to the classical formula above. In the case of an operator such as $-i\hbar \, \partial/\partial x$, the differentiation is understood to apply to the first ψ only and not to $\psi\psi^*$.[14] The product $A_{(op)}\psi\psi^*$ is more normally written $\psi^*A_{(op)}\psi$, to underline the Hermitian character of $A_{(op)}$, and the most common use of the above result is to find an average energy $\bar{E}$ by the use of the formula

$$\bar{E} = \frac{\int \psi^*H\psi\,d\tau}{\int \psi^*\psi\,d\tau} \tag{1.11}$$

This formula is of particular value in calculations on polyelectronic atoms and molecules where no exact wavefunction is calculable. It answers the question 'What would be the energy of the system if the electrons were forced into having a distribution represented by a hypothetical wavefunction ψ?'

1.11   Exact measurements

Let us now consider the special case where ψ is an eigenfunction of $A_{(op)}$. Thus $A_{(op)} = a\psi$, and thus

$$\frac{\int \psi^*A_{(op)}\psi\,d\tau}{\int \psi^*\psi\,d\tau} = a \tag{1.12}$$

Thus a is the special case of $\bar{A}$ when the 'average' is taken over a sharply defined property. In particular, the eigenfunctions of H describe the states of the system which have sharply defined energy. These will from now on be described as the energy levels of the system, or in the case of an atom as the atomic orbitals. In order to discover the eigenfunctions of H one must solve the differential equation

$$H\psi = \frac{-\hbar^2}{2m}\left(\frac{\partial^2}{\partial x^2} + \frac{\partial^2}{\partial y^2} + \frac{\partial^2}{\partial z^2}\right)\psi + V\psi = \varepsilon\psi \tag{1.13}$$

This equation is known as the Schrödinger equation. Each solution of it corresponds to a certain energy ε, and not all energies are possible. If the potential energy, V, varies with time *no* solutions are possible, and in fact the equation is strictly speaking invalid. However, another more general equation is available[15] to give us some information about such a system. This may all be seen as a consequence of the relation $\Delta\varepsilon\,\Delta t \geqslant h/4\pi$. A time variation of V amounts to a partial defining of t, and results in lack of definition of ε.

[14] $\psi\psi^*$ is not even a proper *wave*function as it can be independent of time.
[15] See chapter 4.

1.12 Quantization and normal modes

The fact that not all values of ε are possible in a given system explains the existence of discrete energy levels, and is thus a major triumph of quantum theory. This 'quantization' is a general consequence in any system described by a differential equation whose solutions have fixed values at more than one point,[16] or which must repeat at regular intervals. Let us consider in detail one such differential equation, that for a string of uniform tension T and mass per unit length ρ, stretched along the x-axis but distendable in the y-direction, and fixed at the points $(0, 0)$ and $(L, 0)$[17] (see figure 1.8). The solution of this equation illustrates several principles used in spectroscopy. A mechanical analysis of any small section of the string shows that the following relationship must always hold:

$$\frac{\partial^2 y}{\partial t^2} = \frac{T}{\rho} \frac{\partial^2 y}{\partial x^2} \tag{1.14}$$

This is an example of a classical wave equation, relating a time differential to a space differential. It is obeyed by a very large number of possible

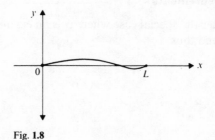

Fig. **1.8**

contortions of the string, and although these contortions can all be shown to fit the equation, they cannot all be directly derived from it. They may, however, be derived indirectly, as will be shown.

Certain solutions can be derived directly. These are motions in which every part of the string moves in phase (or 180° out of phase) with every other part of the string. Such motions are known as 'normal modes'. This is illustrated in figure 1.9. The algebraic version of the definition of a normal mode is a solution of the equation of type $y = F(x) G(t)$ where $F(x)$ is independent of t and $G(t)$ independent of x. Because $F(x)$ is independent of time, the profile of the string is unchanged except in sign or overall vertical scale during the vibration.

We may now solve the differential equation directly under the assumption that our solution will be a normal mode. By substituting $y = F(x) G(t)$ we obtain

$$F(x) \frac{d^2 G(t)}{dt^2} = \frac{T}{\rho} G(t) \frac{d^2 F(x)}{dx^2} \tag{1.15}$$

[16] In the case of wavefunctions $\psi(x, y, z)$ the 'points' are $\pm \infty$, where $\psi = 0$.

[17] There are particularly clear discussions of this problem in W. Kauzmann, *Quantum Chemistry*, Academic Press, New York, 1957, chs 2 and 3; and in G. Stephenson, *Mathematical Methods For Science Students*, Longmans Green, London, 1961, ch. 24.

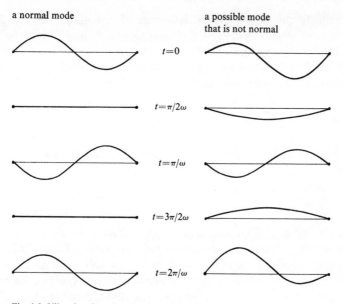

a normal mode   a possible mode
                                           that is not normal

$t=0$

$t=\pi/2\omega$

$t=\pi/\omega$

$t=3\pi/2\omega$

$t=2\pi/\omega$

Fig. **1.9** Vibrational modes of stretched string
(vertically exaggerated)

and hence

$$\frac{1}{G(t)}\frac{d^2G(t)}{dt^2} = \frac{T}{\rho}\cdot\frac{1}{F(x)}\frac{d^2F(x)}{dx^2} \tag{1.16}$$

Note that the substitution allows us to replace partial differentials by ordinary ones. Now the left-hand side of 1.16 is quite independent of x, and the right-hand side is quite independent of t. They can thus only remain equal for all x and t if both are, in fact, constant. Thus we may extract two new equations,

$$\frac{d^2G(t)}{dt^2} = -k^2G(t) \tag{1.17}$$

and

$$\frac{d^2F(x)}{dt^2} = \frac{-k^2\rho F(x)}{T} \tag{1.18}$$

where we have set both sides of 1.16 equal to the constant $-k^2$, chosen for convenience. This technique of extracting several ordinary differential equations from one partial differential equation is known as 'separation of variables', and is only possible for relatively simple partial differential equations. Because of this it is often only possible to solve Schrödinger's equation by making various simplifications in the physical model of the atom or molecule concerned.

The only possible solutions of equation 1.17 are $G(t) = A\sin(kt+\phi)$, where A and ϕ are constants. The frequency of the oscillation is $2\pi/k$, and ϕ merely determines the phase of the oscillation at any given time.

Similarly the only possible solutions of equation 1.18 are $F(x) = C \sin k\sqrt{\rho/T}\,x + D \cos k\sqrt{\rho/T}\,x$. The constants C and D may be determined from the way the string is stretched initially. We know two facts about this. Firstly, when $x = 0$, $F(x) = 0$. This proves that $D = 0$. Secondly, when $x = L$, $F(x) = 0$. Substituting,

$$0 = C \sin k \sqrt{\frac{\rho}{T}} \cdot L \tag{1.19}$$

Therefore either $C = 0$ (i.e. there is no motion at all) or else

$$k = \frac{n\pi}{L} \sqrt{\frac{T}{\rho}}$$

where n is integral. This second solution is possible because $\sin n\pi = 0$.

Thus, because of our boundary conditions $F(0) = 0$ and $F(L) = 0$, we are forced to conclude that k may not have any value, but may only have certain values which in this case are evenly spaced. And as k determines the frequency of the vibration, we conclude that normal modes only exist with certain fixed frequencies. Our final solution is

$$y = F(x).G(t) = AC \sin \frac{n\pi x}{L} \sin \left(\frac{n\pi t}{L} \sqrt{\frac{T}{\rho}} + \phi \right) \tag{1.20}$$

The normal mode illustrated in figure 1.9 is in fact the mode for which $n = 2$.

These normal modes are sometimes also described as standing waves. Of course, they only 'stand' in that their profile $F(x)$ is independent of time. Their amplitude varies sinusoidally in proportion to $G(t)$. Very similar solutions are obtained for the classical problem of the oscillation of air in a tube and for the quantum mechanical problem of a particle or photon in a one-dimensional box (see section 1.3).

Schrödinger's equation (1.13) is in fact the x, y, and z part of a more general wave equation to be discussed in chapter 4. It is derived from the more general equation by a separation of time and space variables along exactly the same lines as above. This explains why H must be independent of time in order for Schrödinger's equation to be soluble; if it is not, the separation of variables is not possible. The time-dependent part of the solutions of the wave equation where H is independent of time ('steady state solutions') is

$$G(t) = e^{-i\varepsilon t/\hbar} \tag{1.21}$$

where ε is the energy of the state and $\varepsilon/\hbar$ corresponds to the frequency of the wave motion associated with the particle. Thus an analogy may be drawn between atomic orbitals and classical normal modes, or 'standing waves'. In both cases, of course, it is not the amplitude that is fixed but the space profile of the amplitude. This is not to say that the electron density in an atom necessarily fluctuates with time. For the electron density is proportional to $\psi^*\psi$. If $\psi = F$ (space coordinates). $e^{-i\varepsilon t/\hbar}$ as it must in an atomic orbital, then $\psi^*\psi = F^2$ (space coordinates), which is independent of time.

1.13 Other modes

So far we have only discovered those solutions of equation 1.14 corresponding to normal modes. However, all other possible motions of the string are in fact describable as combinations of normal modes. This is a consequence of some very sophisticated mathematical reasoning well beyond the scope of this book. Fourier theory, mentioned in section 1.6, is a closely related consequence of this reasoning. It may be seen that the second, non-normal vibration illustrated in figure 1.9 may be analysed as an out-of-phase mixture of the fundamental mode ($n = 1$) and the first harmonic ($n = 2$).

The above principle applies not only to sinusoidal modes but to any set of normal modes which are solutions of a differential equation. In particular, the solutions of the Schrödinger equation form such a set, and therefore any conceivable wavefunction ψ (subject to the same boundary conditions) may always be expressed as a series in terms of the known atomic orbitals ψ_i resulting from any solution of the Schrödinger equation.

$$\psi = \sum_i a_i \psi_i \tag{1.22}$$

This fact is of considerable value in dealing with the effects of small potential energy changes ('perturbations') on known systems. The above idea of being able to describe any mode of oscillation in terms of normal modes is also important in the analysis of molecular motions. Thus any vibration of a molecule may be analysed into contributions from a limited number of normal modes whose frequencies (or combinations thereof) are measured by infrared and Raman spectroscopy. The continuous string in our example above might be considered as a special case of a 'molecule' with an infinite number of atoms and hence, even though movement is restricted to one dimension, an infinite number of normal modes.

Suggestions for further reading

M. Jammer, *The Conceptual Development of Quantum Mechanics*, McGraw-Hill, New York, 1966.
M. Karplus and R. Porter, *Atoms and Molecules*, Benjamin, New York, 1970.
P. W. Atkins, *Molecular Quantum Mechanics*, OUP, London, 1970, vols I and II.
R. P. Feynman, R. B. Leighton, and M. Sands, *Lectures on Physics*, Addison-Wesley, Reading, Mass., 1963.
M. W. Hanna, *Quantum Mechanics in Chemistry*, Benjamin, New York, 1969.

Problems [18]

1.1   The visible emission spectrum of excited hydrogen atoms consists of lines at 6546, 4869, 4379, and 4102 Å. Devise a graphical method to determine the series to which these lines belong, and to derive R_∞ in units of cm^{-1}.

1.2   The basic quantization condition underlying the Bohr theory was that $\int pdq = nh$, where p is the momentum of the system, q is its space coordinate and n is integral.

Consider the circular Bohr orbits for the hydrogen atom (i.e. p constant). If the force between the electron (mass m) and the nucleus (mass effectively infinite) is $e^2/4\pi\varepsilon_0 r^2$, show that the allowed energies are $-R_\infty/n^2$, where $R_\infty = e^4 m/8\varepsilon_0^2 h^3 c$.

1.3   Using Planck's radiation equation (1.3), prove Wien's displacement law, $v_{\text{max. intensity}} \propto T$.

1.4   The F_2 molecule has vibrationally excited states whose energy separations from the ground state are $892n$ cm^{-1}, where n is integral. Plot an approximate graph of the variation of the average vibrational energy of one molecule of F_2 from $T = 0$ to $T = 500$ K, and, from this graph, deduce the variation of vibrational specific heat with temperature.

1.5   Using the data in problem 1.2, and assuming that $\Delta p \simeq 2p$ and $\Delta r \simeq 2r$, show that the uncertainty principle applied to Bohr's model predicts a ground state electronic wavefunction consistent in extent with Bohr's predicted orbit. (Use the relation kinetic energy $= -\frac{1}{2} \times$ potential energy, which holds for a central Coulomb force field and is known in its general form as the Virial Theorem).

1.6   What is the uncertainty in position of a helium atom (mass 6.6×10^{-27} kg) in liquid helium, where the mean speed of the atoms is about 100 m s^{-1}? Comment on the result.

1.7   If gas molecules collide on average every 10^{-12} seconds in such a way that their rotation is affected, calculate the minimum linewidth for any spectroscopic transition involving rotational energy, in cm^{-1} and Hz.

1.8   An electron microscope is to be capable of resolving points 0.1 μm apart. Using the principle mentioned in section 1.8, calculate the minimum voltage through which the electrons must be accelerated.

1.9   **a** A particle of mass m is confined to move around a ring of radius r so that its position is described by an angle θ. If its momentum is constant, and its potential energy zero, derive a classical expression for its energy and hence show that the wave equation

$$\frac{\hbar^2}{2mr^2} \frac{\partial^2 \psi}{\partial \theta^2} + \varepsilon\psi = 0$$

is appropriate.

b Solve this equation to find the allowed energies and wavefunctions of the particle. How do these energies depend upon r?

The above system is a simplified model for an electron confined to an aromatic ring. How does it help to explain the observation that benzene and naphthalene are colourless whereas many polycyclic hydrocarbons are coloured?

1.10   A particle of mass m is confined within a hard walled one-dimensional box

[18] See section 5.1 for energy units.

of length l, such that its potential energy is zero within the box and infinite outside it.

What is the wave equation for the particle within the box? What are the boundary conditions on the wavefunction? Solve the wave equation with these boundary conditions to find the allowed wavefunctions of the particle. Hence calculate the lowest possible energy jump that the particle can make.

This system is a simplified model for a proton trapped in a metal lattice. The transition mentioned above may be detected by the resonant absorption of slow neutrons whose de Broglie wavelength corresponds to the transition energy. The proton and neutron both have a mass of 1.67×10^{-27} kg. Calculate the velocity of the resonant neutrons if the box is 0.05 nm in length.

1.11 A free particle of momentum p, moving in the x-direction, has a wavefunction $\psi = e^{-i(kx + \omega t)}$, which represents a plane wave of wavelength $2\pi/k$ and frequency $\omega/2\pi$. Show that this wavefunction is an eigenfunction of the momentum operator $-i\hbar \partial/\partial x$, and that de Broglie's relation holds.

1.12 A string is free to vibrate in any direction normal to the line joining the points between which it is stretched. How will the locus of the midpoint of the string in the lowest vibrational state depend upon the phase angles ϕ_1 and ϕ_2 (equation 1.20) for the motion analysed along two normal axes? What higher energy states are possible?

1.13 The first two excited levels of the fluorine atom are 404.0 cm^{-1} (doubly degenerate) and 102 406.5 cm^{-1} (six-fold degenerate) above the four-fold degenerate ground state. Determine the fraction of atoms in these excited levels at 1000 K relative to the ground state. Hence calculate the total heat (per mole) absorbed by the electron motion at this temperature.

2 Solutions of the Schrödinger equation

2.1 One particle, one dimension

Because of a variety of available mathematical techniques, most of which were developed in the eighteenth and nineteenth centuries, it is not usually too difficult to solve the Schrödinger equation (1.13) for a single particle with a single degree of freedom, for a wide variety of possible potentials, V. One of the simplest solutions occurs when one particle has zero potential energy within a certain range of motion, and an infinitely large potential energy outside that range (see problem 1.10). Not surprisingly the solution shows that the particle has no chance of appearing outside its range of zero potential energy. Inside that range its possible wavefunctions are exactly the same as the classical wavefunctions (i.e. profile functions) for the vibrating string. The allowed kinetic energies are proportional to n^2, where n is an integer not less than one. The absence of a state with $n = 0$ is important, for if it existed then a particle could be at rest (i.e. with no uncertainty in velocity) within a confined space (i.e. with limited uncertainty in position), which would contradict the uncertainty principle. The 'particle in a box' solutions give acceptable approximations for the energies of outer electrons in linear conjugated molecules, and help to explain why long conjugated molecules (such as carotene) are coloured.

A related problem to that of the particle in a box is the 'harmonic oscillator', or particle in a parabolic potential well, where $V \propto x^2$. This potential energy function is a good approximation to the actual potential energy variation of a nucleus in a molecule when it is moved short distances in a given direction from its equilibrium position ($x = 0$). The resulting wavefunctions (which in this case give the probability of finding the *nucleus* at a given distance from its equilibrium position) are roughly similar to the 'particle in a box' wavefunctions, although they tail away gradually towards $\pm\infty$ because the walls of the potential well are no longer infinitely steep. Another consequence of the parabolic well is that the permitted energy levels are now evenly spaced rather than diverging as n^2.[1] Once again, zero energies are not permitted, and this gives rise to the phenomenon of vibrational 'zero point energy'. The consequences of all this are discussed more fully in chapter 9.

[1] In a real molecule the potential well widens out more rapidly than a parabola on the side where the bond is stretched. In this case the energy levels converge towards the limit of the dissociation energy.

2.2 One particle, three dimensions

The most important application of the Schrödinger equation is to the case of a particle free to move in three dimensions but attracted by a Coulomb potential $V = -k/r$, where r is the distance of the particle (in any direction) from a central point, and zero potential energy is defined as the energy at very large r. This case is a good approximation to the hydrogen atom, although it treats the nucleus as being infinitely heavy, and spinless, whereas in fact it has a mass approximately two thousand times that of the electron, and possesses a spin magnetic moment. For the hydrogen atom, $k = e^2/4\pi\varepsilon_0$, where e is the magnitude of the electron and nuclear charge and ε_0 the permittivity of free space.

Thus, the equation to be solved in order to find the allowed energies E and wavefunctions ψ of the hydrogen atom is, from equation 1.13,

$$-\frac{\hbar^2}{2m_e}\left\{\frac{\partial^2\psi}{\partial x^2}+\frac{\partial^2\psi}{\partial y^2}+\frac{\partial^2\psi}{\partial z^2}\right\}-\frac{e^2}{4\pi\varepsilon_0 r}\psi = E\psi \qquad (2.1)$$

here m_e is the electron mass.

The solutions of this equation are of enormous significance because of three facts:

i   They form a complete set[2], as discussed in section 1.13, and therefore all other related wavefunctions may be written in terms of them.

ii The results are approximately valid for all atoms, in so far as each electron may be considered as moving in a Coulomb-like potential due to the combined attraction of the nucleus and repulsion of the other electrons. This is a particularly good approximation when the orbit of the electron under consideration lies largely outside the orbits of all the other electrons. Even for electron orbits where this is not so, the results give a good account of the angular distribution of the electron density, because the net non-central forces are always weak. This point is discussed further in section 2.6 (iii).

iii The solutions are closely related to the solutions for other quantum mechanical problems in three dimensions, notably those of the rigid rotor and the nuclear spin moment.

The solution is described in detail in almost all quantum mechanics textbooks, and the following is only an outline designed to bring out points of particular relevance to the theory of spectroscopy. The first stage in the solution is to separate the variables. However, it is not possible to separate the variables x, y, and z because of the potential energy term $e^2/4\pi\varepsilon_0 r$. The term r is equal to $\sqrt{x^2+y^2+z^2}$, and this square root cannot be reduced to parts depending upon x, y, or z alone. To avoid this difficulty, it is necessary to change the Cartesian coordinates to spherical coordinates r, θ, and ϕ (see figure 1.5) by making the three substitutions:

$$x = r\sin\theta\cos\phi,\ \ y = r\sin\theta\sin\phi,\ \ z = r\cos\theta \qquad (2.2)$$

[2] This is only strictly true when the very high-energy, unbound ($E > 0$) states are included.

The corresponding transformation of the Schrödinger equation is a famous and extended exercise in the manipulation of partial differentials. Equation 2.1 finally becomes:

$$\frac{\hbar^2}{2m_e}\left\{\frac{1}{r^2}\frac{\partial}{\partial r}\left(r^2\frac{\partial\psi}{\partial r}\right)\right\} + \frac{\hbar^2}{2m_e r^2\sin\theta}\left\{\frac{\partial}{\partial\theta}\left(\sin\theta\frac{\partial\psi}{\partial\theta}\right)\right\}$$
$$+ \frac{\hbar^2}{2m_e r^2\sin^2\theta}\frac{\partial^2\psi}{\partial\phi^2} + \left(\frac{e^2}{4\pi\varepsilon_0 r} + E\right)\psi = 0 \tag{2.3}$$

If we now assume that $\psi = \psi(r).\psi(\theta).\psi(\phi)$, where $\psi(r)$ depends upon r only, and so forth,[3] we may apply the technique of separation of variables as described in section 1.12. The details of the separation form problem 2.1. First one separates functions of r from those of θ and ϕ, and arrives at the two equations:

$$\frac{\hbar^2}{2m_e}\left\{\frac{1}{r^2}\frac{d}{dr}\left(r^2\frac{d\psi(r)}{dr}\right)\right\} + \left\{\frac{e^2}{4\pi\varepsilon_0 r} + E - \frac{\hbar^2 l(l+1)}{2m_e r^2}\right\}\psi(r) = 0 \tag{2.4}$$

and

$$\frac{\psi(\phi)}{\sin\theta}\left\{\frac{d}{d\theta}\left(\sin\theta\frac{d\psi(\theta)}{d\theta}\right)\right\} + \frac{\psi(\theta)}{\sin^2\theta}\left\{\frac{d^2\psi(\phi)}{d\phi^2}\right\} + l(l+1)\psi(\theta)\psi(\phi) = 0 \tag{2.5}$$

where $\hbar^2 l(l+1)/2m_e$ is merely a convenient way of writing the separation constant without, so far, any further implications. However, a comparison of equation 2.5 with table 1.2 shows that 2.5 may be rewritten:

$$L^2_{(op)}(\psi(\theta)\psi(\phi)) = \hbar^2 l(l+1)\psi(\theta)\psi(\phi)$$

where $L^2_{(op)}$ represents the operator corresponding to a measurement of the square of the total angular momentum. Thus the solutions of equation 2.5, in addition to describing the angular distribution of the electronic wavefunction in the hydrogen atom, also provide solutions for the angular distribution and orbital angular momentum of any particle about the centre of any direction-independent field. They also show that an electron in any one atomic orbital of hydrogen will have a total angular momentum whose magnitude (but not direction) is sharply defined. The combined solutions $\psi(\theta)\psi(\phi)$ are known as 'spherical harmonics', and occur in many branches of physics. They are commonly written in the notation $S_{lm}(\theta, \phi)$.

Equation 2.5 may be further split into the two equations:

$$\frac{1}{\sin\theta}\left\{\frac{d}{d\theta}\left(\sin\theta\frac{d\psi(\theta)}{d\theta}\right)\right\} + \left\{l(l+1) - \frac{m^2}{\sin^2\theta}\right\}\psi(\theta) = 0 \tag{2.6}$$

and

$$\frac{d^2\psi(\phi)}{d\phi^2} + m^2\psi(\phi) = 0 \tag{2.7}$$

[3] 'ψ' is simply used here to signify a wavefunction or part of a wavefunction. It will generally be true that different ψ's, such as $\psi(r)$, $\psi(\theta)$, and $\psi(\phi)$ will be quite different and distinct algebraic functions.

Here the separation constant is m^2, and equation 2.7 is obviously very similar to equation 1.17, the solutions being $\psi(\phi) = A \sin(m\phi + \text{constant})$. It is convenient to write these in the alternative form $\psi(\phi) = Be^{im\phi}$.

Because $\psi(\phi)$ must repeat every time ϕ increases by 2π (see example 1.9), it follows that m must be integral. m is thus a quantum number for the hydrogen atom. Also, $Be^{im\phi}$ is an eigenfunction of the operator $-ih\,\partial/\partial\phi$ which was stated in table 1.2 to be the operator $L_{z(\text{op})}$ corresponding to a measurement of the z-component of angular momentum. The corresponding eigenvalue is mh. From this we deduce that the z-component of orbital angular momentum of the electron is a measurable quantity which is restricted to integral multiples of $\hbar$.

Equation 2.6 is much harder to solve than is equation 2.7. Just as m was found to be integral when the condition $\psi(\phi) = \psi(\phi + 2\pi)$ was applied so also l is found to be integral when the conditions that $\psi(\theta)$ is finite is applied. The solutions obtained are different for each separate integral value of l and of m, and no solutions at all are possible if $m > l$ or if $m < -l$. In other words, $-l < m < l$. Some of the best known eigenfunctions are:

l	m	$\psi(\theta)\psi(\phi)$	Conventional cartesian description
0	0	1	s
1	1	$\sin\theta\, e^{i\phi}$	$p_x + ip_y$
1	0	$\cos\theta$	p_z
1	-1	$\sin\theta\, e^{-i\phi}$	$p_x - ip_y$
2	0	$3\cos^2\theta - 1$	d_z^2

A moment's inspection shows that the two orbitals $l = 1$, $m = \pm 1$ are the same functions of θ and ϕ as the wavefunctions of the usual p_x and p_y orbitals $[\psi(\theta)\psi(\phi) = \sin\theta\cos\phi$ and $\sin\theta\sin\phi$ respectively] taken in the combinations $p_x \pm ip_y$.

Equation 2.4 is just as hard to solve in the general case as is equation 2.5, for the solutions depend upon the actual values of l (though not of m). In other words, the radial distribution of the electron density is linked to its angular distribution. The eigenvalues of the $\psi(r)$ do not, however, depend on l in this simple one-electron case, but only on the 'principal' quantum number n. Solutions of equation 2.4 are only possible provided n is a positive integer, and also provided that $n > l$. This is because of the boundary condition that $\psi(r)$ must fade to zero as r approaches infinity. Hence l is confined to the values $n-1, n-2, \ldots 0$.

The overall, total wavefunctions $\psi = \psi(r)\psi(\theta)\psi(\phi)$ are the solutions of the original equation 2.3, and their eigenvalues give the allowed orbital energies E of the hydrogen atom, namely

$$E = \frac{-m_e e^4}{8h^2\varepsilon_0^2 n^2} \tag{2.8}$$

This formula (after small corrections for the finite nuclear mass and spin, and for relativity) gives excellent agreement with the measured values of R_∞, the Rydberg constant. The wavefunctions themselves are conventionally described as 1s, 3d, 4f, etc., where the number gives the value of n and the letter the values of l, according to the code s (for $l = 0$), p (for $l = 1$), d, f, g, h, etc.

2.3 The rigid rotor

The radial solutions $\psi(r)$ are unique to hydrogen-like atoms and ions, although they do not require much modification in order to describe the orbital of a single, loosely bound electron outside an inner core of nucleus plus tightly bound electrons. However, the angular functions or spherical harmonics, $\psi(\theta)\psi(\phi)$, have much more general application. This is because they are independent of the attractive potential V provided that the force arising from the latter is always directed towards the centre of coordinates. Thus they are also the correct wavefunctions for the rigid linear rotor and indeed for all simple angular momentum problems. A rigid linear rotor, for example, is a system of particles each of which is constrained by an (idealized) potential well of infinite depth which rules out its radial motion completely. Because this rotor differs from the hydrogen atom only in the radial force field, the wavefunctions of each particle, and hence of the rotor as a whole, may be described like the hydrogen orbitals, as s-orbital like, p-orbital like, etc. Furthermore, they describe states of the particle in which the total value of the angular momentum and its z-component are sharply defined, because the $\psi(\theta)\psi(\phi)$ are eigenfunctions not only of $H_{(op)}$ but also of $L^2_{(op)}$ and of $L_{z(op)}$, with eigenvalues $l(l+1)\hbar^2$ and $m\hbar$ respectively.

Now it will be noticed that the angular momentum eigenfunctions, unlike those of $H_{(op)}$, in which a radial part is also included, are quite independent of the particle masses or positions. In other words, what we have discovered about the angular momentum of the electron holds true for the angular momentum of all systems in which the particles lie on a straight line. (A further analysis also can be made to show that very similar quantization rules exist for the angular momentum of bodies in general, although complications arise because of the extra possible axis of rotation.) Furthermore, we can work backwards from the known allowed angular momenta in order to find the allowed kinetic energies of rotation.

Let us illustrate this by taking the CO molecule as an example, and assuming (in this approximation) that it is effectively rigid. This molecule has a moment of inertia I about its centre of mass which may be calculated from the known nuclear masses m and their distances from the centre of gravity r, using the formula $I = \Sigma m r^2$, and is, in fact, $1.457 \times 10^{-46} \, \mathrm{kg \, m^2}$. Now we know from the above reasoning that its angular momentum is constrained to be $\hbar\sqrt{l(l+1)}$ (direction unspecified), where l is a non-negative integer. We also know that its angular momentum is $I\,d\theta/dt$, where θ is the angle that the molecule has rotated about its chosen rotation axis, and also that the kinetic energy $E = \frac{1}{2}I(d\theta/dt)^2$. Hence, $E =$ (angular momentum)$^2/2I$; that is,

$$E = \frac{\hbar^2}{2I}\sqrt{l(l+1)} \tag{2.9}$$

This formula gives the allowed energies of the rigid rotor. It is more normally written with the letter J in place of the letter l, and will be subsequently quoted in that form.

It is instructive to consider the detailed rotational wavefunctions $\psi(\theta)\psi(\phi)$. Firstly, when J (or l) $= 0$ the angular momentum and all its components are sharply defined as being zero. Hence, as we would expect, $\psi(\theta)\psi(\phi)$ is s-orbital like, which is a way of saying that the direction of the molecular axis in space is totally undefined, as predicted in section 1.8 (iii). When $J = 1$, the magnitude of the total angular momentum is precisely $\sqrt{2}\hbar$, because $J^2_{(op)}\psi = J(J+1)\hbar^2\psi$. Also its x and y components remain quite unknown and its z component must be $\pm\hbar$ or 0. The zero z component means that the nuclear wavefunction is like a p_z-orbital, with the average direction of the molecular axis being along the z-axis. The $\pm\hbar$ components belong to the orbitals of type $p_x \pm ip_y$, which indicate that the molecular axis is biased towards (but not necessarily in) the xy plane. There are thus three possible rotational states, distinguished by their z component of angular momentum, for $J = 1$. In general, rotational states of quantum number J have degeneracies of $2J+1$.

2.4   Spin

Reference has already been made to the Stern and Gerlach experiment (section 1.7), in which a beam of silver atoms was split into two components with equal and opposite deflections by an inhomogeneous magnetic field. Now the unpaired electron in the silver atom is in an s-orbital, and hence has no orbital angular momentum. Hence, this experiment showed that the electron has an intrinsic angular momentum whatever its physical state. Furthermore, the production of just two beams, and their observed splitting, showed that this intrinsic angular momentum, or 'spin', had a fixed value of $\sqrt{3}\hbar/2$, and z-components of $\pm\frac{1}{2}\hbar$, as if l were allowed to be $\frac{1}{2}$. This most unexpected finding shows that electron 'spin' is a fundamentally new property of matter and not just due to rotation of a particle of finite dimensions; the latter would be bound only to have integral values of l, like all ordinary rotating bodies. 'Spin' is a special property of certain fundamental particles, notably electrons, protons, and neutrons, all of which have exactly the same intrinsic angular momentum of $\sqrt{3}\hbar/2$, i.e. 'a spin of $\frac{1}{2}$'.

There is an interesting complementary relationship between spin angular momentum, with its half-integral quantum numbers, and orbital angular momentum, with its integral quantum numbers l and m which we discussed in the previous two sections. In sections 1.8 (ii) and (iii) we considered the implications of the uncertainty principle for angular momentum measurements in general. If these implications are fully pursued, then with the aid of the operator transformations given in table 1.2, it is possible to deduce a great deal about the rigorous quantum mechanical limitations that are imposed upon all total angular momentum measurements. The line of argument, although complex, is strictly logical, and is one of the most pleasing and elegant parts of quantum mechanics.[4] From it one concludes, without needing to calculate any wavefunctions, that the only possible total angular momenta that any particle or body can have are $\hbar\sqrt{j(j+1)}$, where j is either zero or a positive integer or exactly half a positive integer. Furthermore, one discovers that only one of the components of the total angular momentum is measurable (conventionally the z component) and that this may only take the values $\hbar m_j$, where m_j is either an integer or a half-integer whose allowed values go in steps of 1 from j to $-j$. The arguments that prove this hold true whatever interactions there may be between the internal angular motions that contribute to the total angular momentum $\hbar\sqrt{j(j+1)}$.

Now this discovery is perfectly consistent with what we have discovered from the wave equation about orbital angular momentum. But it also leaves room for the possibility of half-integral quantum numbers— a possibility that is realized by the existence of spin. Thus, although spin is a fundamental property of particles whose origins have yet to be revealed by nuclear physics, nevertheless because it has the characteristics of an angular momentum, it is bound by the strict laws, derived from the uncertainty principle, that govern all angular momenta.

For most purposes the spin orientation (i.e. z component) of a particle may be considered as an independent variable of its motion. The limitations of this simplification are described in section 2.6 (i) and also (for the case of interacting nuclear spins) in section 7.9.

2.5 Two particles, three dimensions

We have already dealt with two problems involving more than one particle by making approximations, namely, fixing the nucleus in the hydrogen atom and fixing the bond length in the rigid rotor. Had we not made these approximations, we would have had to consider both particles together. Suppose that in general we have two particles a and b. The combined wavefunction may be written $\psi(a, b)$ and the Schrödinger equation, derived as before, is

$$-\frac{\hbar^2}{2m_a}\nabla_a^2\psi(a, b) - \frac{\hbar^2}{2m_b}\nabla_b^2\psi(a, b) + V\psi(a, b) = E_{a,b}\psi(a, b) \qquad (2.10)$$

where $E_{a,b}$ is the combined energy of particles a and b and ∇_a^2, ∇_b^2 refer to the coordinates of a and b respectively. If the two particles do not

[4] See, for example, P. W. Atkins, op. cit., chapter 6.

interact, then this equation is simply a conflation of the separate Schrödinger equations for a and b, and may be separated into these by assuming that $\psi(a, b) = \psi(a)\psi(b)$ (i.e. that the total probability density is the product of the individual densities). The individual particle wavefunctions can then be separated into r, θ, and ϕ parts as before.

However, we are interested in cases where the two particles interact strongly, such as in a diatomic molecule.[5] In such cases it is sometimes possible to separate the variables and solve the combined Schrödinger equation by a change of coordinate system.[6] When it is possible, then instead of considering the separate cartesian or polar coordinates of a and b, we consider the three cartesian coordinates X, Y, and Z of the centre of gravity together with a coordinate R describing the orientation of the interparticle axis relative to the cartesian axes.

This coordinate system has enormous advantages, because

a the motion of the centre of gravity of a free body such as a molecule is independent of all motions of the constituent particles which leave the centre of gravity of the body fixed. The resulting translational wavefunction $\psi(XYZ)$ is simply that of a free point particle whose mass equals the molecular mass. Thus, three of our six degrees of freedom are easily accounted for.

b in simple cases the potential energy V only affects the interparticle separation R, and is independent of θ and ϕ. Thus, the complete wave equation can be solved exactly as before except that the interparticle potential will not normally be Coulombic and thus the radial part of the wave equation may present difficulties.

We can put

$$\psi(a, b) = \psi(X)\psi(Y)\psi(Z)\psi(R)\psi(\theta)\psi(\phi) \tag{2.11}$$

and then calculate the translational rotational and vibrational wavefunctions separately. When the conditions above hold true, it can be shown (problem 2.3) from equation 2.10 that:

$$E_{a,b} = E_X + E_Y + E_Z + E_R + E_{\theta + \phi} \tag{2.12}$$

or in other words

$$E_{total} = E_{translation} + E_{vibration} + E_{rotation}$$

provided that the various potential and kinetic energies of the molecule are functions of X, Y, Z, R, and θ and ϕ alone. Of course the latter condition is usually only approximately true. For example, if the molecule rotates rapidly this will add an extra 'centrifugal' term to the internuclear potential. Also, the vibrational motion will affect the moment of inertia

---

[5] It is also possible to deal with the hydrogen atom by this method, without having to assume that the nucleus is fixed.

[6] Indeed, provided the interparticle force always lies along the interparticle axis, it is *always* possible to separate the motion of the centre of mass of the system from those mutual motions of the particles which leave their centre of mass fixed.

by changing the bond length. However, in the next section the latter complication will be shown not to matter and the former complication can be shown to necessitate only a small correction to the wavefunctions and energies that are calculated by ignoring it. Thus the phrases 'rotational energy' and 'vibrational energy' have considerable meaning, even when the effects of rotation and vibration are not fully separable.

2.6   Many particles, further dimensions

The variables describing many-particle systems can always be separated in the same way as those describing two-particle systems if a suitable coordinate system can be found to separate out different non-interacting types of motion. However, the motions are rarely completely non-interacting and therefore we must consider certain important approximations more closely.

(i)   Separation of spin motion

The spin of a particle may be described as belonging to an extra dimension in so far as it does not interact with other particle motions. In such cases the 'spin coordinates' may be separated from all the spatial coordinates. The total wavefunction is then simply the product of the spin wavefunction (which is commonly written in the code form 'α' or 'β', meaning $s = \frac{1}{2}$ and $m_s = +\frac{1}{2}$ or $-\frac{1}{2}$) multiplied by the ordinary spatial wavefunction. When more than one particle of any one kind is involved, a further quantum mechanical law, Pauli's exclusion principle, comes into operation to limit the allowed combinations of the multi-particle spin and space functions. This is discussed in chapter 8 in the context of one of its important applications.

As before, weak interactions between spin and other electron and nuclear motions may be treated as minor perturbations requiring only small corrections to be made to the simple separate-coordinate wavefunctions. Such interactions include hyperfine coupling and the chemical shift (chapter 7). However, there is a much stronger interaction which occurs between the magnetic moments of unpaired electron spins and their orbital motion in states with $l \geqslant 1$, in which the electron also possesses orbital angular momentum and hence an orbital magnetic moment. This interaction is known as spin-orbit coupling, and it arises from the tendency of the spin magnetic moment and the orbital magnetic moment to align themselves mutually. The effects of this on the atomic wavefunctions are complex, and in general it is no longer possible to identify distinct spin and orbital angular momentum states, whereas if there had been no spin orbit interaction, we would have been able to identify a spin angular momentum of $\hbar\sqrt{s(s+1)}$ (s being equal to $\frac{1}{2}$ for a single electron) and an orbital angular momentum of $\hbar\sqrt{l(l+1)}$.

Fortunately, however, the total angular momentum resulting from the combination of the spin and orbital angular momenta, whatever their mutual interaction, still obeys the fundamental laws outlined in section 2.4. In other words, it is restricted to values of $\hbar\sqrt{j(j+1)}$ and to z components of $m_j\hbar$, where $j > m_j > -j$ and j is a positive integer or half-integer. Furthermore, it is obvious from the law of conservation of angular momentum that $\hbar\sqrt{j(j+1)}$ cannot exceed $\hbar\sqrt{l(l+1)}+\hbar\sqrt{s(s+1)}$ nor be less than $|\hbar\sqrt{l(l+1)}-\hbar\sqrt{s(s+1)}|$.[7] This can be shown (problem 2.4) to mean that $l+s > j > |l-s|$. Hence, we are still able to make at least a partial prediction of the total magnetic moment of single electrons (and, by implication, of electrons in combination) in atomic states. Various rules (e.g. Hund's rules) are available to predict which of the possible values of j represent the lowest energy state, and the origin of these rules is discussed in chapter 10.

(ii)   Separation of nuclear and electronic spatial motion

The interaction between electrons and nuclei is very strong, and at first sight it would seem impossible to separate the spatial coordinates of the two types of particle. However, an approximate separation is possible because of the fact that any nucleus has a mass of at least 1836 times that of the electron. The detailed justification of this was first made by Born and Oppenheimer in 1927, and the approximate coordinate separation is named after them. Their argument is a very complex one, involving perturbation theory, but fortunately one can become convinced about their conclusions by accepting the intuitively reasonable idea that the nuclei vibrate (i.e. move relative to each other) much more slowly than the electrons orbit, because of their greater mass.

This idea has two important consequences. Firstly, the nuclei are effectively static during the time in which the electron makes several orbits. This is the justification for treating the nuclei as fixed when calculating electronic wavefunctions. Secondly, however strongly the electrons may pull any one nucleus, they are bound to move around it many times before any effect is observed. Hence, the nucleus is only sensitive to the average pull of the electrons, and not to their instantaneous pulls, rather in the way that large particles do not exhibit detectable Brownian motion but do move with bulk movements of the fluid in which they are suspended. This is the justification of the way in which a molecular bond in the previous section was considered simply as a steady binding potential whereas it is basically made up of a number of Coulombic attractions and repulsions between point charges.

The above conclusions, in a more precise form, justify the use of the concept of independent molecular electronic and nuclear wavefunctions, the overall wavefunction being the product of these.

---

[7] $|z|$ means the modulus, or positive value of z.

(iii) Separation of internuclear motions

The Born-Oppenheimer argument may be extended to deal with a rotating molecule. One argues intuitively that the nuclear rotational motion is much slower than both the electronic *and* the nuclear vibrational motion. In other words, a molecule vibrates many times during any rotation, and hence the moment of inertia of the molecule is determined only by the mean internuclear separation. Thus, the rotational potential and kinetic energies are effectively independent of the vibrational motions, so that the Schrödinger equation for the nuclei can be separated into rotational and vibrational parts. Such a separation was described in section 2.5 for the diatomic molecule; however, Born and Oppenheimer's approximation achieves the separation for all molecules.

The remaining nuclear motions in a polyatomic molecule are all of vibrational type. Provided that they are of small amplitude they may be further separated into normal modes, as defined at the end of chapter 1. The number of normal modes needed to describe them is easily calculated. For an n-atomic molecule it is $3n-6$, or $3n-5$ if the molecule is linear. This is because n free particles have $3n$ degrees of freedom, none of which can be removed by any interparticle potential. A non-linear molecule has three translational and three rotational degrees of freedom, leaving $3n-6$ vibrational degrees, and a linear molecule has three translational and two rotational degrees of freedom, there being no freedom for the particles to vary their zero angular momentum about the linear axis.

One may crudely state that normal vibrational modes are normal (i.e. independent) because they are all of some different type, or else of the same type but in directions mutually at right angles. Thus, the four

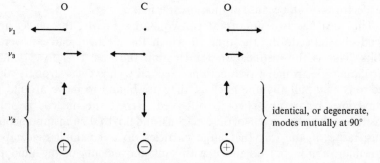

Fig. **2.1** Normal vibrational modes of CO_2

normal vibrational modes of carbon dioxide are as in figure 2.1. In the symmetrical stretching mode, v_1, only the oxygen atoms move. In the asymmetrical stretching mode, v_3, the oxygen-oxygen distance is fixed and the main movement is that of the carbon atom. All bending of the molecule is accounted for by the doubly degenerate mode v_2, which can be thought of as a single mode which accounts for two degrees of freedom because movement can occur in any direction in the plane at right angles to the original molecular axis. The motions v_1 and v_3 can each be described by a single coordinate and v_2 can be described by two coordinates. These four coordinates form a 'complete set'. In other words, although many other more complex vibrations of the carbon dioxide molecule can occur, any and every one of these complex vibrations may be analysed into four components, one from each normal mode.

An important feature of normal modes is their mutual independence; so far we have only justified this by a crude argument about different types of motion. In order for the modes to be strictly independent, it is necessary that when the vibrational part of the Schrödinger equation for the nuclei is transformed into the normal vibrational coordinates, it is then separable into one equation for each coordinate. This will only occur if, following the coordinate transformation, the various potential energies for stretching and bending of bonds fall into separate terms, each involving only one normal coordinate (as is obvious from section 1.12). The condition for this to occur is, in fact, that all the basic bond stretching and bending potentials must be harmonic. In other words, the restoring forces in each case must be proportional to the distortions from equilibrium (Hooke's law). The mathematical reasoning behind this requirement is slightly complex (see problem 2.5), but the point of it may be seen by considering the asymmetrical CO_2 stretching vibration v_3. If the individual C—O bonds obey Hooke's law, then the restoring force on either oxygen atom will be the same, however the carbon atom moves, and hence the oxygen atoms will never move relative to each other. However, if the C—O bonds do not obey Hooke's law (either separately or as a result of combination), then the oxygen atom cannot move precisely in phase, and hence v_3 will not be strictly normal to other modes.

Real molecular bonds only obey Hooke's law approximately, at low vibrational amplitudes. In other words, their oscillations are slightly anharmonic. The physical effects of anharmonicity are discussed in section 9.8. Fortunately, they may be considered as minor additions to the behaviour that is expected, and in the main observed, with harmonic oscillators.

If the Schrödinger equation can be separated into equations each involving one normal vibrational coordinate with a harmonic potential energy term, then it follows that each normal mode of motion will have its own independent energy levels. Also, because it is independent of other motions, it may be thought of as having its own classical vibration frequency which will, in general, differ from the frequency of any other normal mode.

In conclusion, we can see that provided certain approximations are made, it is possible to analyse all the possible motions of the nuclei and electrons in most molecules into 'electronic', 'vibrational', 'rotational', and 'spin' categories, and further to subdivide the vibrational motions into normal modes. From now on this book will consider these categories separately.

Suggestions for further reading

As in chapter 1, plus:

I. N. Levine, *Quantum Chemistry*, Allyn and Bacon, Boston, 1970.

J. M. Anderson, *Introduction to Quantum Chemistry*, Benjamin, New York, 1969.

Problems

2.1   **a**   By putting $\psi = \psi(r)\psi(\theta, \phi)$ in equation 2.3, show that the equation may be completely divided into terms dependent upon r alone and upon θ and ϕ but not upon r. Hence, using a separation constant of $\hbar^2 l(l+1)/2m_e$, derive equation 2.4.

   b   By similarly putting $\psi(\theta, \phi) = \psi(\theta)\psi(\phi)$, derive equation 2.5, and by using a separation constant of m^2, derive equations 2.6 and 2.7.

2.2   Show that the wavefunctions $\psi = 1$, $\psi = \sin\theta\, e^{i\phi}$ and $\psi = 3\cos^2\theta - 1$ are all eigenfunctions of the operator $L^2_{(op)}$ (table 1.2) with eigenvalues of 0, $2\hbar^2$, and $6\hbar^2$ respectively and also of $L_{z(op)}$ with eigenvalues of 0, $\hbar$, and 0 respectively.

2.3   If equation 2.10 is transformed into centre of mass coordinates X, Y, Z, R, θ, and ϕ, then it may be separated, using equation 2.11, into six equations involving only X, Y, Z, R, θ, and ϕ respectively, provided that the potential energy of the system depends only upon R and that the moment of inertia is effectively constant in time. Show that, under these conditions, the total energy of the system is simply the sum of the energies of each separate type of motion (equation 2.12).

2.4   Prove that if $\sqrt{l(l+1)} + \sqrt{s(s+1)} \geqslant \sqrt{j(j+1)} \geqslant \sqrt{l(l+1)} - \sqrt{s(s+1)}$ where $l \geqslant s$ and l, s, and j are positive integers or half-integers, then $l+s \geqslant j \geqslant l-s$. Also show that in fact $\sqrt{j(j+1)}$ can never reach the limits of $\sqrt{l(l+1)} \pm \sqrt{s(s+1)}$ and hence that two mutually interacting angular momenta can never set exactly parallel or antiparallel to each other.

2.5   Consider a linear carbon dioxide molecule in which the oxygen(1), carbon and oxygen(2) nuclei have coordinates x_1, x_2, x_3 respectively relative to some arbitrary origin on the line of the molecular axis, and in which the equilibrium C—O bond length is r. The potential energy V of the nuclei in their vibrations along the molecular axis will be

$$V = \tfrac{1}{2}k\left[(r - x_2 + x_1)^2 + (r - x_3 + x_2)^2\right] \quad \text{(where } k \text{ is a force constant)}$$

provided that the individual C—O bonds obey Hooke's law both separately and in combination.

   a   Justify this expression for the potential energy.

   b   Show that it is possible to write down three new independent coordinates q_1, q_2, and q_3, such that q_1 describes the position of the molecular centre of gravity, q_2 the sum of the separations of the oxygen nuclei from the centre of gravity, and q_3 the difference of these separations, and that

$$q_1 = \frac{(x_1 + x_3)m_O + x_2 m_C}{2m_O + m_C}, \quad q_2 = x_3 - x_1$$

and

$$q_3 = \frac{(x_1 + x_3 - 2x_2)m_C}{2m_O + m_C}$$

where m_O and m_C are the masses of the oxygen and carbon nuclei respectively.

c Show that q_2 and q_3 describe distortions of the CO_2 molecule from equilibrium according to the normal modes v_1 and v_3 respectively (see figure 2.1).

d Show that the potential energy V, written in these new coordinates q_1, q_2, and q_3, becomes

$$V = \tfrac{1}{2}k\left[2\left(r - \frac{q_2}{2}\right)^2 + \left((2m_O + m_C)\frac{q_3}{2m_C}\right)^2 \right]$$

and hence that the Schrödinger equation for the linear vibrational motion of CO_2 is divisible into three independent equations. Also show that this separation would not have been possible if the individual bonds had not obeyed Hooke's law.

2.6 Show that an electron in a state with $l = 1$, $m = 1$, has no preferred direction in the xy plane. Show also that if there is one electron in each of the states $l = 1$, $m = 1, 0$ and -1, then the total electron density has a spherical distribution.

What can you deduce from this about the tendency of electrons in degenerate orbitals to remain unpaired, as far as is possible?

2.7 The (unnormalized) wavefunctions for the 2s and 2p orbitals of the hydrogen atom are $[(r/a)-2]\,e^{-r/2a}$ and $(r/a)\,e^{-r/2a}\cos\theta$ respectively, where $a = \varepsilon_0 h^2/\pi m_e e^2$. Show that these are both solutions of equation 2.3 with $E = m_e e^4/32 h^2 \varepsilon_0^2$.

3 Classical theory of interaction of radiation and matter

3.1 Electromagnetic radiation

The full physical theory of electromagnetic radiation was developed in the last century by James Clerk Maxwell. His theory is a development of Faraday's experimental observations on the production of magnetic fields by electric currents and of electric fields (e.m.f.'s) by changing magnetic fields. Maxwell showed that these processes did not fundamentally depend on the presence of electrical conductors. He then considered the problem of what happens when an electric field fluctuates (e.g. the field produced by an oscillating electric charge). The fluctuating electric field produces a fluctuating magnetic field, which in turn produces a further fluctuating electric field, and so forth. A particularly interesting fluctuating electric field is one which extends along a line through space and has an amplitude which varies sinusoidally along that line, and whose peaks move with a velocity c along the line. Maxwell showed that, provided c was equal to the product of two constants known from static electromagnetism, a steady state was bound to be set up by such a field (even if it only existed for a brief moment) because it would produce a similarly fluctuating magnetic field exactly at right angles to it, along the same line, which would in turn give rise to the original fluctuating electric field, and so forth. Such a self-supporting system of electric and magnetic fields, illustrated in figure 3.1, constitutes a beam of plane polarized electromagnetic radiation. If some different fluctuating electric field is considered it can be shown that the fields to which it gives rise may be analysed into fields which fluctuate but do not move through space, plus this electromagnetic radiation moving at velocity c. This is analogous to the way in which any disturbance at the surface of a pond produces local

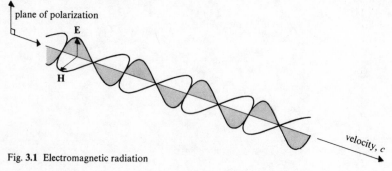

Fig. **3.1** Electromagnetic radiation

40

turbulence (which dies away rapidly if the disturbance ceases) plus steady, sinusoidal outward ripples whose velocity is determined by the properties of water rather than by the original disturbance.

Maxwell crowned his triumph by showing that the predicted velocity c was equal, within experimental error, to the known velocity of light. His relationship between electrostatic constants, electromagnetic constants, and the velocity of light is so important that it has enabled scientists to produce simple SI units to cover the whole of physics, instead of having to use different conflicting units for electrostatics and for magnetism. His theory is a masterpiece of neat and imaginative reasoning and its only drawback is that it is rather hard for non-physicists to follow.[1]

A particularly important source of radiation is the dipole oscillator. This basically consists of two opposite electrical charges whose separation x varies sinusoidally at a frequency of v, so that $x = x_0 \sin 2\pi vt$. The fields produced by these moving charges consist of statically fluctuating electric and magnetic components plus electromagnetic radiation flowing

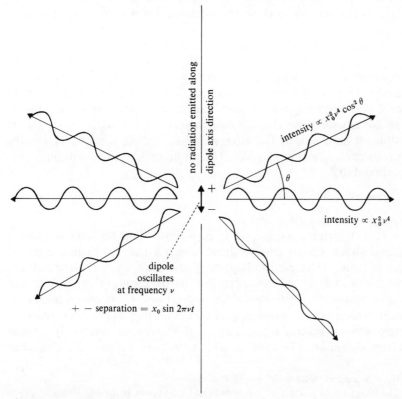

Fig. **3.2** The dipole oscillator

[1] A fairly straightforward account is given by A. F. Kip, *Electricity and Magnetism*, McGraw-Hill, New York, 1962.

outwards from the dipole, mainly in the plane normal to the dipole axis, and plane polarized in the direction of that axis.

The amplitude of the radiation produced is proportional to the acceleration of the charges, i.e. to v^2,[2] and its intensity is proportional to the square of its amplitude,[3] i.e. to v^4. This is illustrated in figure 3.2.

3.2   Classical model of the atom

In the late nineteenth century the existence of the nucleus was unknown, and the atom was thought of as a number of electrons buried in some sort of positively charged jelly. In the free, undisturbed atom the net centre of negative charge coincided with the net centre of positive charge so that the atom had no dipole moment. However, when the atom was put into an electric field the positive and negative charges tended to separate so as to counteract the field. This explained the existence of atomic polarizability, which contributes to the dielectric behaviour of substances. If the electrons were set into motion they were assumed to oscillate sinusoidally with some characteristic frequency or frequencies, v, and gradually to lose their oscillation amplitude due to damping forces until they once again came to rest.

This classical model appears odd to the modern scientist, accustomed as he is to electrons confined by atomic orbitals around a tiny nucleus. However, it is not quite as incorrect as it first appears. For when one studies the effect of a weak electric field E on an atom obeying quantum laws, one discovers that the resultant small distortions of the atomic orbitals amount to an electron motion very like that which was assumed in the classical model. The atom may be quantum mechanical, but the perturbation is almost classical. The resulting dipole moment, μ, is proportional to E:

$$\mu = \alpha E \tag{3.1}$$

where α is the polarizability.

It is, therefore, not surprising to discover that the classical model allows what it was designed to allow, namely a quite far reaching explanation of classical electrical and optical phenomena. These explanations are worthy of study by the spectroscopist, partly because they provide him with a means of visualizing certain interactions whose quantum mechanical explanation is complex, partly because they provide an adequate explanation of several important optical phenomena, and partly because a large collection of bodies (e.g. nuclei) often obeys macroscopic classical

---

[2] If $x = x_0 \sin 2\pi v t$ then $d^2x/dt^2 = -4\pi^2 v^2 x_0 \sin 2\pi v t$.
[3] The energy stored in any field is proportional to the square of the field strength. This is most easily seen by considering the electrical potential energy of the charges in a capacitor. Both the potential of each electron and the number of electrons on each plate are proportional to the voltage V across the capacitor (i.e. to the electric field) and hence the potential energy is proportional to V^2.

laws even when the individual bodies are limited by the more restrictive requirements of quantum mechanics.

3.3   Absorption of light

We have already seen how radiation consists of fluctuating electric and magnetic fields. It follows from this that when the classical atom is subjected to radiation the positive and negative charges in it will be set into forced oscillation at the frequency of the radiation, because they are being constantly pulled apart first one way and then the other by the oscillating electric field of the radiation.

This forced motion takes place against the 'damping forces' which constantly absorb energy. (These forces are now realized to represent the loss of energy to other modes of motion in the atom and beyond, and also via re-radiation.) The larger the amplitude of forced motion, the more energy is lost via damping. The maximum amplitude of forced motion occurs at resonance, when the forcing frequency equals the natural oscillation frequency. Hence the maximum absorption of radiation occurs at certain specific frequencies, which are the natural oscillation frequencies of the electrons. This explains the existence of optical absorption spectra without obvious recourse to quantum mechanics, and also, when done quantitatively, gives a remarkably good description of the observed shape of spectral absorption lines (see chapter 6).

The above paragraphs rely heavily on the theory of forced, damped oscillations. The student who is moderately familiar with differential equations should attempt problem 3.2 which leads through this theory. Less mathematically inclined students may care to consider figure 3.3 which describes the results of an experiment that is quite easy to set up, where a rotating disc and long length of weak elastic provide a sinusoidally varying horizontal force to the end of a rigid pendulum so as to set it into forced oscillation. The phase relationships that are described will be used to explain refraction in a later section.

Another form of oscillating dipole that can absorb radiation is a polar molecule, where the electric charges are permanently partially separated. The dipole of such a molecule can oscillate relative to the electric field of radiation either by changes of bond length or simply by rotation. Thus the molecule is able to take up both vibrational and rotational energy from the radiation, the vibrational energy being at a fixed frequency.

3.4   Scattering and reflection of light

When the classical atom is set into forced oscillation it acts as a dipole oscillator, and emits radiation outwards into space, as shown in figure 3.2, whose frequency is that of the original beam of radiation. This phenomenon is known as Rayleigh scattering. It is central to parts of chapter 9, and is also useful for understanding several aspects of magnetic resonance

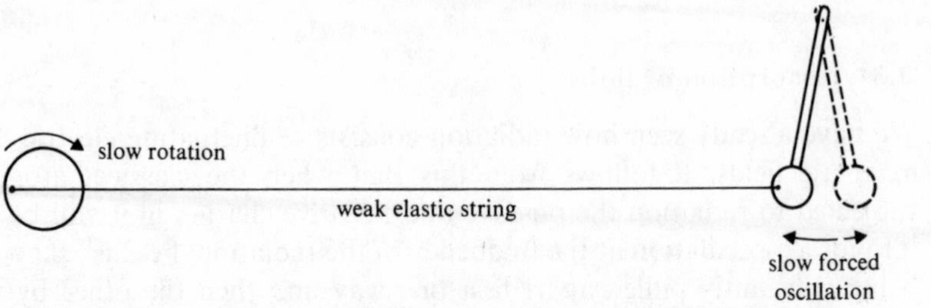

slow rotation

weak elastic string

slow forced
oscillation

(a) The pendulum would like to oscillate faster than the driving force is oscillating. Therefore, the driving force has to provide its maximum pull when the pendulum is fully to the left, in order to force an increase in the oscillation time of the pendulum. Therefore the rotation and oscillation are in phase, i.e. swing together from left to right and back. Most of the work of the driving force is done to kill the natural swing of the pendulum and therefore this swing remains small. The pendulum is forced to follow the rotation for there is no other way in which it can regularly gain any energy at all.

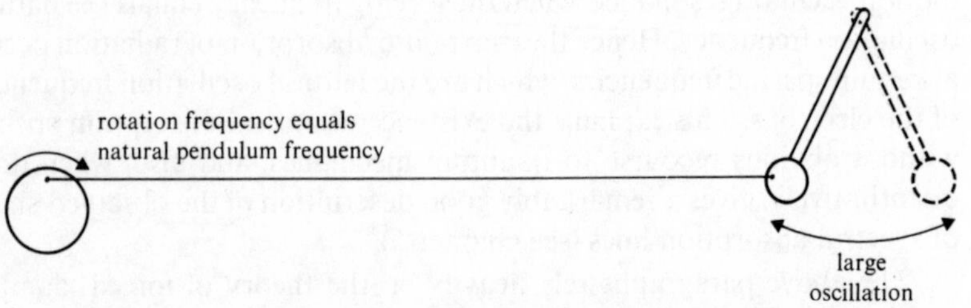

rotation frequency equals
natural pendulum frequency

large
oscillation

(b) The driving force oscillates at the natural frequency of the pendulum. It builds up pendulum oscillations progressively by providing maximum pull every time the pendulum is at the bottom of its swing moving towards the left. Thus the driving force is 90° of phase ahead of the pendulum swing. The limit on the amplitude achieved is set by the damping forces, which absorb energy in proportion to the size and speed of swing. Hence resonance and maximum energy absorption occur at the same frequency.

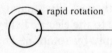

rapid rotation

rapid forced
oscillation

(c) The driving force oscillates much faster than the natural frequency of the pendulum. It provides its maximum pull when the pendulum is over to the right, in order to curtail the natural swing and thus speed up the pendulum, i.e. it is 180° out of phase. The forced swing of the pendulum is once again small and thus not much work is done against the damping forces.

Fig. 3.3 Forced, damped oscillations.

spectroscopy, although in this latter case magnetic rather than electric dipoles and fields are involved, and the induced motion is a circular precession rather than a linear oscillation.

The existence of a bright sky in daytime is due to scattering of the sun's rays by the atmosphere; one can readily observe, using polarizing sunglasses, that the light from a clear sky is polarized in a manner that can be deduced from the fact that the dipole oscillator can only emit radiation polarized in the direction of the dipole axis. The same polarization provides useful information in Raman spectroscopy (section 9.7). The blue colour of a clear sky arises because the efficiency of scattering by small particles is, as we have seen, proportional to the fourth power of the frequency of the light being scattered.

When a row of scattering atoms or ions, or molecules, is considered, as at the surface of a crystal, the spherical waves produced by the atoms combine (interfere) to produce two plane waves, one parallel to the original wavefront of the incident beam and one which constitutes a reflected beam (figure 3.4).

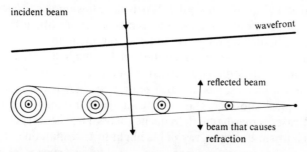

Fig. **3.4** Reflection due to scattering by one layer of atoms

A more detailed analysis of reflection shows that, for surfaces large compared with the wavelength of the incident radiation, interference effects exactly cancel the fourth-power frequency dependence of atomic scattering, and therefore that reflected light has the same colour proportions as the incident light (e.g. white clouds).

3.5 Refraction [4]

The forward scattered beam is exactly parallel to the incident beam. However, the forced oscillation that causes scattering will not in general be exactly in phase with the incident radiation. For example, most atoms and molecules absorb radiation in the near ultraviolet region of the spectrum. Therefore, when they are irradiated with visible light they are forced into oscillations which lag slightly in phase behind the incident light—a situation somewhere between case (a) and case (b) in figure 3.3. This has no observable effect on the reflected beam, but it does affect the

[4] See R. Feynman, *Lectures on Physics*, op. cit., vol. 1, ch. 31.

net beam that carries on into the material. This net beam is a combination of the incident beam, slightly reduced in amplitude, together with the parallel scattered beam, somewhat behind in phase. The combination of these is a wave which lags very slightly in phase relative to the incident beam (figure 3.5).

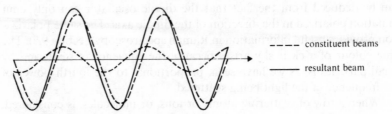

---- constituent beams

—— resultant beam

Fig. **3.5** Combination of two constituent beams with the same frequency but different phase

Each successive layer of atoms produces a similar phase lag, and the net effect of this is an apparent slowing of the wave velocity whilst the beam is passing through the solid. When this slowing is considered on a macroscopic scale one can make the usual deductions about the angle of refraction of light. The refractive index of the medium is, of course, the reciprocal of the fraction by which the speed of the light is reduced in the medium, and is thus directly related to the phase lag of the scattering. This phase lag increases as an absorption frequency is approached, and thus for typical substances absorbing in the ultraviolet, the refractive index increases somewhat with frequency across the visible spectrum. The ability of a prism to bend a ray of blue light further than one of red light is known as dispersion, and provides a well-known means of analysing radiation at optical or near-optical frequencies.

This explanation of refraction shows that substances tend to be very efficient scatterers of radiation near to their absorption frequencies. This explains why metals are usually grey and also are good reflectors. There are so many electron levels in metals (classically, so many possible modes of electron oscillation) that metals are efficient absorbers and scatterers for almost all frequencies of electromagnetic radiation. The theory of reflection and refraction can be extended to give a quantitative explanation of a very wide range of optical phenomena, such as total internal reflection, dichroism, and the polarization of reflected light.[5]

Suggestions for further reading

A. F. Kip. *Electricity and Magnetism*, McGraw-Hill, New York, 1962.
W. Kauzmann, *Quantum Chemistry*, Academic Press, New York, 1957, ch. 15.

[5] The variation of optical rotatory power with frequency (optical rotatory dispersion) and the related absorption effects (circular dichroism) are becoming of increasing importance to chemists. See D. C. Pasto and C. R. Johnson, *Organic Structure Determination*, Prentice-Hall, Englewood Cliffs. N.J., 1969, ch. 7; and W. Kauzmann, op. cit., chs 15 and 16.

R. P. Feynman et al., op. cit., vol. I.

G. Stephenson, *Mathematics for Science Students*, Longmans London, 1961, p. 367 et seq.

Problems

3.1 In his analysis of electromagnetism, Maxwell discovered the following two equations governing the relation between a fluctuating, non-uniform electric field E in the xy plane and a fluctuating non-uniform magnetic field H in the xz plane:

$$\frac{\partial H}{\partial x} = -\varepsilon_0 \frac{\partial E}{\partial t} \quad \text{and} \quad \frac{\partial E}{\partial x} = -\mu_0 \frac{\partial H}{\partial t}$$

In these equations, ε_0 and μ_0 are electromagnetic constants. Show that the only possible steady state solution of these equations is:

$$H = a \sin(vt - vx/c + \phi)$$
$$E = a\sqrt{\mu_0/\varepsilon_0} \sin(vt - vx/c + \phi)$$

where a and ϕ are constants and $1/c^2 = \mu_0\varepsilon_0$. Show also that these solutions describe an electromagnetic wave of velocity $-c$, and that they are valid for all values of the frequency v.

3.2 An oscillating particle of mass m, moving in the x-direction only, experiences three forces: a restoring force $-mk^2x$ which tends to bring it back to its equilibrium point $x = 0$, a periodic driving force of $am \cos \omega t$, and a damping force of $-\lambda m\, dx/dt$, i.e. proportional to its velocity.

a Show that the particle's motion is governed by the equation:

$$\frac{d^2x}{dt^2} + \lambda \frac{dx}{dt} + k^2 x = a \cos \omega t$$

b Show that a solution of this equation (in fact the most general one) is:

$$x = A \exp(-\lambda + \sqrt{\lambda^2 - 4k^2})t/2 + B \exp(-\lambda - \sqrt{\lambda^2 - 4k^2})t/2 +$$
$$+ \frac{a\lambda\omega \sin \omega t + a(k^2 - \omega^2) \cos \omega t}{(k^2 - \omega^2)^2 + \lambda^2\omega^2}$$

where A and B are constants.

c Show that, when t is very large, the first two terms become negligible. (It may help to note that $k \gg \lambda$.)

d Show that the third term accurately describes the experimental observations outlined in figure 3.3, including the phase relation of the forced oscillation to the driving force $a \sin \omega t$.

e The rate of absorption of energy, per cycle, by the damping forces is

$$\int_{\substack{\text{one} \\ \text{cycle}}} \lambda \frac{dx}{dt} dx$$

Show that this may also be written as

$$\int_0^{2\pi/\omega} \lambda \left(\frac{dx}{dt}\right)^2 dt$$

and hence that the rate of absorption of energy varies with the driving frequency ω according to the expression

$$\frac{\lambda \pi a^2 \omega}{(k^2 - \omega^2)^2 + \lambda^2 \omega^2}$$

(This explains why the out-of-phase component of the oscillation is sometimes called the 'absorption' component).

Show also that, provided $\lambda \ll k$, this describes a reasonable shape for a spectral absorption line (figure 4.2).

3.3   A man wearing polarizing sunglasses studies the sky vertically above him at sunset on a clear evening, and discovers that the light from the sky is almost completely polarized. Explain this, by considering the unpolarized light from the sun to consist of equal components of horizontally and vertically polarized light. In what direction is this scattered light polarized?

3.4   Calculate the wavelength of radiation with a frequency of 100 MHz (a typical television frequency). How does this compare with the normal dimensions of a television aerial? Suggest a reason for this, remembering that an electrical impulse cannot pass down a conductor at a speed faster than that of light.

3.5   Explain why satisfactory infrared and visible absorption spectra can be obtained by analysing the radiation reflected from the surface of solids. To what extent does this technique study (a) surface molecules, (b) molecules near the surface, (c) molecules deep within the solid?

3.6   Give an approximate explanation of why unpolarized light, when reflected other than straight back by a surface, becomes at least partially polarized. Should the polarizing lenses in sunglasses used by car drivers have their polarizing axes oriented horizontally or vertically in order to reduce glare from wet roads?

3.7   When a beam of unpolarized light is refracted by a calcite crystal it is observed to split into two beams, one of which is polarized at right angles to the other. What can you deduce from this about the structure of calcite, if you already know that calcite is composed of Ca^{2+} ions and planar CO_3^{2-} ions in which each oxygen atom is equivalent?

4 Quantum mechanical theory of interaction of radiation and matter

4.1 Introduction

The classical explanation of the interaction of radiation and matter is able to predict the existence of spectral absorption lines and their shapes, and to explain many optical phenomena in terms of scattering. It is, however, quite unable to predict the actual frequencies of absorption, the actual linewidths, or the intensities, because of the inadequacies of the classical model of the atom. Our quantum mechanical theory must rectify this without significantly changing the classical conclusions about scattering.

We have so far only considered wavefunctions for systems in which the potential energy V does not vary with time. In such cases we were able to solve the equation $H\psi = E\psi$ in order to calculate ψ in terms of the space coordinates of the system. A word of explanation is now necessary to explain why the Schrödinger equation is only soluble when V does not vary with time. In the previous three chapters we have considered the function ψ in terms of its dependence on space and spin coordinates although the original definition (section 1.9) and the word 'wavefunction' imply a dependence of ψ upon time. For example, the un-normalized wavefunction of the $1s$ orbital of the hydrogen atom (ignoring spin) was stated to be $\psi = e^{-r/a}$. In fact the full wavefunction is $\psi = e^{-r/a}.e^{-i\varepsilon t/\hbar}$, where ε is the energy of the $1s$ orbital. The extra, time-dependent term does not affect the equation $H\psi = E\psi$ provided that V does not contain t. If it does, then ψ is no longer an eigenfunction of H and the Schrödinger equation is almost useless. Only the average values of properties may be determined, and the states are said to be no longer 'sharply defined' or 'stationary'. The extra term $e^{-i\varepsilon t/\hbar}$ represents a sinusoidal fluctuation of ψ, and it makes ψ closely analogous to the normal modes of vibration of classical systems discussed in section 1.12. It also helps to explain the importance of normalization, for when such a 'stationary state' wavefunction is multiplied by its complex conjugate the result is independent of time. If $\psi = e^{-r/a}e^{-i\varepsilon t/\hbar}$, then $\psi^* = e^{-r/a}e^{+i\varepsilon t/\hbar}$, and so $\psi\psi^* = e^{-2r/a}$.

4.2 The (time-dependent) wave equation [1]

Although sharply defined quantum states do not exist in systems whose potential energy fluctuates with time, there is still a great deal that can

[1] Of course, this is a tautology. All wave equations are by definition time-dependent.

49

be discovered about such systems, which is fortunate for the understanding of spectroscopy. In chapter 1 we derived Schrödinger's equation from the assumption that the average value of a measurement, A, on a system in the normalized state, ψ, was $\int \psi^* A_{(op)} \psi \, d\tau$, where $A_{(op)}$ was derived from the classical variable to be measured by substituting $-i\hbar \partial/\partial x$ for p_x, and leaving x as it stood. Suppose that we continue to use this assumption and now apply Newton's second law of motion to the system, i.e.,

force = mass × acceleration.

This may be rewritten, according to classical Hamiltonian mechanics, as

$$-\left(\frac{\partial H}{\partial x}\right)_p = \frac{\partial p}{\partial t} \tag{4.1}$$

For convenience we work in one dimension only. Note that if H (the classical Hamiltonian) = KE + PE, then because the KE only depends on p, $(\partial H/\partial x)_p = (\partial PE/\partial x)_p$ which is the negative of the force acting on the particle.

Let us consider the measurement of $\partial p/\partial t$ in quantum mechanics. The operator corresponding to a measurement of p is $-i\hbar(\partial/\partial x)_p$ and hence the corresponding operator for $\partial p/\partial t$ is $-i\hbar(\partial^2/\partial x \, \partial t)_p$, which must be identical (by equation 4.1) with the operator $-(\partial H_{(op)}/\partial x)_p$. Hence we see that $H_{(op)}$ is equal to the operator $+i\hbar \partial/\partial t$, or that

$$i\hbar \frac{\partial \psi}{\partial t} = H_{(op)}\psi \tag{4.2}$$

This is known as the 'time-dependent' wave equation, and holds true whether or not $H_{(op)}$ is time-dependent. Its existence underlines the principle that a full wavefunction, ψ, for a particle contains all possible information not only about the spatial motion of the particle but also about the way this motion changes with time, and hence about its future motion.

If we solve equation 4.2 when $H_{(op)}$ is time-independent, we may use solutions of normal mode type, i.e. $\psi = \psi(\text{space})\psi(t)$. The equation can then be split into two further equations as with previous examples:

$$i\hbar \frac{\partial \psi(t)}{\partial t} = \varepsilon \psi(t) \tag{4.3}$$

and

$$H_{(op)}\psi(\text{space}) = \varepsilon \psi(\text{space}) \tag{4.4}$$

Equation 4.4 is simply the Schrödinger equation, and equation 4.3 is easily solved to give $\psi(t) = e^{-i\varepsilon t/\hbar}$.

The fact that the quantum-mechanical operator for energy may be written as $+i\hbar \partial/\partial t$ shows a pleasing symmetry between energy/time and momentum/distance, which is important in relativity theory, and also helps to explain why the uncertainty principle is valid for uncertainties in energy and time.

4.3 'Transition probabilities'

We now consider just what information of interest to spectroscopy can be extracted from the wave equation, 4.2. Immediately we meet a dilemma, for we would like to discover what we think of as the 'transition probabilities' between stationary states, although we have already admitted that, whilst transitions are occurring (i.e. whilst H is time-dependent), stationary states do not exist!

Fortunately this dilemma can be avoided by returning to the basic quantum mechanical principle of only talking about what one can measure. And what one measures in spectroscopy is, typically, the mean rate of absorption by a macroscopic system of radiation energy of a given frequency. Thus let us suppose that a system has a complex, time-dependent wavefunction ψ. The average energy, $\bar{\varepsilon}$, of the system above that of its ground state, ε_0, will be

$$\bar{\varepsilon} - \varepsilon_0 = \frac{\int \psi^* H \psi \, d\tau}{\int \psi^* \psi \, d\tau} - \varepsilon_0 \tag{4.5}$$

Furthermore the average energy apart from the fluctuating potential energy due to the time-dependent perturbation will be given by 4.5 if H is the unperturbed Hamiltonian. The rate of absorption of energy will be $d(\bar{\varepsilon} - \varepsilon_0)/dt$. Now it will always be possible to write ψ as a linear sum in terms of the orthonormal stationary state space wavefunctions ψ_j that are deduced from the time-independent wave equation for the system. This is not because of any assumption about the states in which the time-dependent system exists, but simply because the time-independent states are a complete set. Thus:

$$\psi = \sum_j a_j \psi_j \tag{4.6}$$

where the coefficients a_j contain all the time-dependence of the system. Substituting 4.6 into 4.5, we obtain:

$$\bar{\varepsilon} - \varepsilon_0 = \frac{\int \sum_j \sum_k a_j^* \psi_j^* H a_k \psi_k \, d\tau}{\int \sum_j \sum_k a_j^* a_k \psi_j^* \psi_k \, d\tau} - \varepsilon_0 \tag{4.7}$$

In this equation, different subscripts j and k are used to indicate that the double sum contains all possible products of the $a_j \psi_j$ with their complex conjugates. Now we know that

$$\int \psi_j^* \psi_k \, d\tau = 0 \text{ if } j \neq k, \quad \text{and} \quad = 1 \text{ if } j = k$$

We also know that $H\psi_k = \varepsilon_k \psi_k$, where ε_k is the energy of state k, provided that H is the unperturbed Hamiltonian, and thus we are content to measure the energy transferred by the perturbation and not the immediate potential energy it produces. Therefore equation 4.7 may be greatly simplified to give:

$$\bar{\varepsilon} - \varepsilon_0 = \frac{\sum\limits_k a_k^* a_k (\varepsilon_k - \varepsilon_0)}{\sum\limits_k a_k^* a_k}$$

Furthermore, if ψ is normalized then $\sum\limits_k a_k^* a_k = 1$ and therefore

$$d(\bar{\varepsilon} - \varepsilon_0)/dt = \sum_k (\varepsilon_k - \varepsilon_0) \, d(a_k^* a_k)/dt \qquad (4.8)$$

We will use this expression to calculate the appearance of spectra, and in doing so we will discover that in many systems either no coefficients other than a_0, or only one coefficient a_k, is of significant magnitude for any one frequency of incident radiation, so that the sum over k is usually not necessary. In such cases the rate of absorption of energy is proportional to $\varepsilon_k - \varepsilon_0$, which is a hint that energy is absorbed in quanta whose magnitude is proportional to $\varepsilon_k - \varepsilon_0$.

What is normally described as the 'transition probability from 0 to k' of a system will be shown to be the integral of $d(\varepsilon_k - \varepsilon_0)/dt$ over the frequency range for which a_k is of significant magnitude, or in other words the total energy absorbed per second from 'white' radiation of unit intensity at all frequencies, by the absorption centred at frequency $(\varepsilon_k - \varepsilon_0)/h$. Although this energy absorption is all that one can accurately talk about, it is nevertheless convenient to talk of systems 'making transitions between states', and this language can be shown to have some approximate validity in most cases.

4.4   Interaction of classical radiation with the hydrogen atom

Let us consider a hydrogen atom, as described by quantum mechanics, whose stationary spatial wavefunctions are written for convenience as ψ_j, and whose actual wavefunction ψ at time t is given by equation 4.6. It is important to remember that the a_j are time-dependent variables which it is our task to investigate. What happens when this atom (whose unperturbed Hamiltonian we write as H_0) interacts with classical radiation?[2] If the field $E_0 \cos \omega t$ of the radiation of frequency ω is in the x

---

[2] Quantized radiation is considered in the final section of this chapter.

direction, and if the x coordinate of the electron of charge e is x, then the extra electrostatic potential energy of the electron will be simply $eE_0x \cos \omega t$ provided that, as in the case with visible and lower frequencies, E_0 is constant over all three dimensions of the molecule. This will mean that the electron will have a lower potential energy on one side of the atom than on the other in the x direction. We wish to discover the effect of this extra potential energy term in the Hamiltonian, which will now be $H_0 + eE_0x \cos \omega t$. In other words we must solve the equation

$$ih \frac{\partial \psi}{\partial t} = (H_0 + eE_0x \cos \omega t)\psi \tag{4.9}$$

Substituting for ψ using equation 4.6, we obtain

$$ih \sum_j \psi_j \frac{da_j}{dt} = \sum_j a_j H_0 \psi_j + \sum_j a_j \psi_j eE_0x \cos \omega t \tag{4.10}$$

Now we multiply through by ψ_k^* and integrate over all space and spin coordinates (ψ_k is another ψ_j, but k does not necessarily equal j):

$$ih \sum_j \frac{da_j}{dt} \int \psi_k^* \psi_j d\tau$$
$$= \sum_j a_j \int \psi_k^* H_0 \psi_j d\tau + \sum_j a_j eE_0 \cos \omega t \int \psi_k^* x \psi_j d\tau \tag{4.11}$$

It is important to note that in the integrals one must include all terms dependent on r, θ, ϕ, and spin, but not on time. Thus $\cos \omega t$ is excluded because it depends on time only, and E_0 is excluded because the electric field of a light wave is effectively constant over the small dimensions of an atom or molecule at any one moment in time. However, x and H_0 may not be excluded.

But we have already chosen ψ_k, ψ_j to be the ordinary space wavefunctions of the hydrogen atom, so that, as before,

$$\int \psi_k^* \psi_j d\tau = 0 \text{ if } j \neq k \quad \text{and} \quad = 1 \text{ if } j = k$$

also

$$H_0 \psi_j = \varepsilon_j \psi_j$$

Therefore we can simplify 4.11 drastically to obtain:

$$ih \frac{da_k}{dt} = a_k \varepsilon_k + \sum_j a_j eE_0 \cos \omega t \, x_{jk} d\tau \tag{4.12}$$

Here x_{jk} is written as shorthand for

$$\int \psi_k^* x \psi_j d\tau$$

Equation 4.12 is really a series of linked differential equations in a series of variables a_j, one of which is a_k. As they stand these equations are not fully soluble, and we therefore have to resort to approximation devices. The most obvious is to consider the rate of change of a_k at the instant at which the time-dependent interaction starts. At this instant all the a_j have their normal stationary state values. For example, if we start in the ground state ψ_0, $a_0 = e^{-i\varepsilon_0 t/\hbar}$, and all the other $a_j = 0$. We would thus find a_k as a function of time, and we could if we wished feed this 'first-order' result back into the original equation 4.12 in place of setting all $a_j = 0$, in order to obtain a better, 'second-order' approximation to the true value of the a_k at all values of t.

At present we will confine ourselves to the 'first-order' theory, by putting all the a_j in 4.12 equal to zero except for a_0, which becomes $e^{-i\varepsilon_0 t/\hbar}$, and a_k itself of course. Now $x_{kk} = 0$, as may be readily seen from a symmetry argument, and therefore we obtain a series of separated equations for the various a_k, namely:

$$i\hbar \frac{da_k}{dt} = a_k \varepsilon_k + x_{0k} e E_0 \cos \omega t \, e^{-i\varepsilon_0 t/\hbar}$$

This is a well known type of differential equation. We solve it by first multiplying through by the integrating factor $e^{i\varepsilon_k t/\hbar}$, obtaining:

$$i\hbar \, e^{i\varepsilon_k t/\hbar} \frac{da_k}{dt} - a_k \varepsilon_k \, e^{i\varepsilon_k t/\hbar} = x_{0k} e E_0 \cos \omega t \, e^{i(\varepsilon_k - \varepsilon_0)t/\hbar}$$

Now the left-hand side of this may be rewritten as:

$$\frac{d}{dt}(i a_k \hbar \, e^{i\varepsilon_k t/\hbar})$$

hence

$$a_k = \frac{1}{i\hbar} e^{-i\varepsilon_k t/\hbar} \int_0^t e^{i(\varepsilon_k - \varepsilon_0)t/\hbar} \cdot x_{0k} e E_0 \cos \omega t \, dt.$$

Thus,

$$a_k^* a_k = \frac{x_{0k}^2 e^2 E_0^2}{\hbar^2} \left| \int_0^t \cos \omega t \cdot e^{i(\varepsilon_k - \varepsilon_0)t/\hbar} \, dt \right|^2 \tag{4.13}$$

Note that $x_{0k} = x_{0k}^*$ because ψ_k, ψ_0, and x are all real.

The next stage in this (quite long) derivation is to evaluate the time integral. First let us make the convenient substitutions

$$\varepsilon_k - \varepsilon_0 = \hbar\omega_k \quad \text{and} \quad \cos \omega t = \tfrac{1}{2}(e^{i\omega t} + e^{-i\omega t})$$

Then the time integral becomes equal to

$$\frac{1}{2} \int_0^t e^{i(\omega_k + \omega)t} \, dt + \frac{1}{2} \int_0^t e^{i(\omega_k - \omega)t} \, dt$$

$$= \frac{1}{2} \left[\frac{e^{i(\omega_k + \omega)t}}{i(\omega_k + \omega)} + \frac{e^{i(\omega_k - \omega)t}}{i(\omega_k - \omega)} \right]_0^t$$

$$= \frac{1}{2i} \left(\frac{e^{i(\omega_k + \omega)t} - 1}{\omega_k + \omega} + \frac{e^{i(\omega_k - \omega)t} - 1}{\omega_k - \omega} \right) \tag{4.14}$$

This contains two separate terms, the former of which is small unless $\omega_k \approx -\omega$ and the latter of which is small unless $\omega_k \approx \omega$. The latter case means $\varepsilon_k - \varepsilon_0 = \hbar\omega$, which is the normal Bohr frequency condition. The first term is normally negligible, except in the case of spectral absorptions whose linewidth is comparable with the complete spectral frequency range. In these latter cases it leads to an unsymmetrical absorption line.

The term in $\omega_k - \omega$, multiplied by its complex conjugate, is

$$\frac{(e^{i(\omega_k-\omega)t}-1)(e^{-i(\omega_k-\omega)t}-1)}{4(\omega_k-\omega)^2}$$

$$= \frac{2-2\cos(\omega_k-\omega)t}{4(\omega_k-\omega)^2} = \frac{\sin^2(\omega_k-\omega)t/2}{(\omega_k-\omega)^2}$$

and therefore:

$$a_k^* a_k = \frac{x_{0k}^2 e^2 E_0^2}{\hbar^2} \cdot \frac{\sin^2(\omega_k-\omega)t/2}{(\omega_k-\omega)^2} \qquad (4.15)$$

at a time t after the radiation is first applied to the system. Now let us suppose that the system stays in a given state, on average, for a time τ, after which it undergoes some sudden physical change that effectively returns it to its initial state ψ_0. This is a fairly realistic model for example for gas molecules undergoing collisions.[3] Between $t = 0$ and $t = \tau$ the total energy absorbed has been (using equation 4.8)

$$\int_0^\tau (\varepsilon_k - \varepsilon_0) \frac{d}{dt}(a_k^* a_k)\, dt$$

and hence the mean rate of absorption per second, $R(\omega)$, of radiation energy of frequency ω, has been

$$R(\omega) = [(\varepsilon_k - \varepsilon_0)a_k^* a_k]_0^\tau / \tau = \frac{\hbar\omega_k}{\tau}[a_k^* a_k]_0^\tau \qquad (4.16)$$

[3] In order to justify this quantum mechanically one has to remember that the full wavefunction for the free atom is $\Psi(X, Y, Z) \times \Psi(R, \theta, \phi, \text{spin})$, where the two parts are separable because the potential energy only affects $\Psi(R, \theta, \phi, \text{spin})$. The centre of mass motion described by $\Psi(X, Y, Z)$ is not affected by radiation, and hence has been allowed to remain hidden in the normalization constant so far in this chapter. But if the extra time-dependent potential due to a collision is considered, the centre of mass motion will no longer be separable from the electron motion during the collision. Furthermore, transitions from excited electronic states with low atomic kinetic energy to ground electronic states with high atomic kinetic energy will be rather likely because of the near-continuum of available high energy translational states, which will make all the frequency components of the collision interaction effective instead of just a selected few. Thus all excited electronic states will become very shortlived during a collision, although the ground electronic state will not, because for the reverse process of collisional excitation to be likely, one would at normal temperatures require the existence of translational states of negative kinetic energy.

It is not strictly necessary on this model for each collision to return the excited atom to its ground state. All that is necessary is that a_0 remain relatively unaffected by the radiation, and that $a_k \psi_k$ be altered substantially upon collision. In fact, a change merely in the phase of a_k is sufficient.

Hence, using equation 4.15, and remembering that $a_k = 0$ at $t = 0$

$$R(\omega) = \frac{\omega_k e^2 E_0^2 x_{0k}^2}{\hbar} \cdot \frac{\sin^2 (\omega_k - \omega)\tau/2}{(\omega_k - \omega)^2 \tau} \tag{4.17}$$

This is our final expression for the transition probability from 0 to k at frequency ω. It contains two particularly interesting terms. The first is the integral

$$ex_{0k} \equiv e \int \psi_k^* x \psi_0 \, d\tau.$$

This is known as the transition moment between states k and 0, and vice versa, and is discussed in section 4.7. The second is the term in $(\omega_k - \omega)$, which gives the line shape, indirectly.

4.5 The line shape function

The expression

$$\frac{\sin^2 (\omega_k - \omega)\tau/2}{(\omega_k - \omega)^2 \tau}$$

is plotted in figure 4.1.

It will be seen that the horizontal scale, and hence the width of the line, is inversely proportional to τ. This tells us that the width of a spectral absorption line is inversely proportional to the lifetime of the absorbing

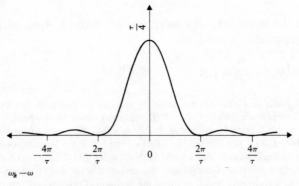

Fig. **4.1** Line shape function for fixed τ

system between external disturbances (such as collisions). We have achieved a more precise derivation of the result discussed in section 1.8 (iv), which now avoids the incorrect implication that when a photon of known energy is absorbed by a system of known energy the resulting energy (as distinct from state) is imprecisely known.

Our line shape function so far does not bear a very strong resemblance to commonly observed line shapes. This is because we have assumed a fixed value for τ. If we recognize that τ will in fact have a range of values with the larger ones becoming decreasingly probable, then we can see that the net line shape resulting from the superimposition of the shape in figure 4.1 for different values of τ will probably lose its oscillations and look more like a typically observed absorption line. In fact, if first order kinetics are assumed for the collision process, so that the probability of the system lasting for a time τ before collision is $e^{-b\tau}$, then the net line shape function may be fairly easily calculated to be

$$1/2\tau_0((\omega_k-\omega)^2+(1/\tau_0)^2)$$ (4.18)

where $\tau_0 = 1/b$ = average lifetime between collisions. This is precisely the same Lorentzian shape as is predicted by the classical theory. It is shown in figure 4.2 in comparison with the Gaussian (Doppler) line shape of section 6.2(ii).

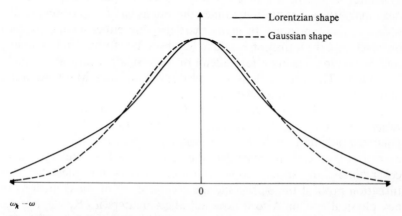

—————— Lorentzian shape

- - - - - - Gaussian shape

$\omega_k - \omega$

Fig. **4.2** Comparison of absorption line shapes

4.6   Total rate of absorption of energy

The total rate of absorption of energy, R, by the transition $k \leftarrow 0$ is the energy that the transition would absorb from 'white' radiation of uniform intensity at all frequencies. If this radiation has an electric field E_0 at all frequencies, then the total energy absorption rate is simply

$$\int_0^\infty R(\omega)\,d\omega \quad \text{or} \quad \int_{-\omega_k}^\infty R(\omega)d(\omega-\omega_k)$$

Hence, from equation 4.17,

$$R = \frac{\omega_k e^2 E_0^2 x_{0k}^2}{\hbar} \cdot \int_{-\omega_k}^\infty \sin^2\frac{(\omega-\omega_k)\tau/2}{(\omega_k-\omega)\tau}\,d(\omega-\omega_k)$$

Provided that τ is not too short, the lower limit may be rewritten as $-\infty$, for absorption only occurs in the frequency range $\omega = 0$ to ∞. The integral between limits $\pm\infty$ is a standard one, and equals $\pi/2$. It is simply the area under the line in figure 4.1. Hence

$$\frac{R}{\hbar\omega_k E_0^2} = \frac{\pi e^2 x_{0k}^2}{2\hbar^2} \tag{4.19}$$

$R/\omega_k E_0^2 \hbar$ is what is commonly termed the transition probability for unit intensity of plane polarized radiation. Equation 4.19 shows it to be independent of line shape, but dependent on the square of the transition moment. This is why peak areas of absorptions with identical transition moments, similar frequencies, but different linewidths may be compared to give the relative proportions of absorbing species in a mixture. It is the basis of quantitative analysis by spectroscopic means.

4.7   The transition moment

Transition moments are of great importance in spectroscopy because their magnitude directly determines the intensity of spectroscopic absorptions and emissions. In many cases they are either identically zero (in which case the transition is said to be 'absolutely forbidden') or almost zero (transition 'strongly forbidden') or very small (transition 'weakly forbidden'). These restrictions are called selection rules. Many examples of their use are quoted in chapters 7 to 10.

The integral $\int \psi_k^* x \psi_0 \, d\tau$ was derived above for a specific physical system. It is the transition moment for electric dipole absorption of x-polarized radiation by the $k \leftarrow 0$ transition of the hydrogen atom. However, it is very easy to generalize the conclusions from this system to cover almost any spectroscopic transition. It does not affect the perturbation proof if the ψ_j become the complete set of wavefunctions of any physical system. Also it does not affect matters if $e E_0 x \cos \omega t$, the energy of interaction with the radiation, is replaced by any other fluctuating interaction. Several such interactions are considered in section 4.9. A very simple example of our more general result is interaction with y- or z-polarized radiation, where the transition moment simply becomes $\int \psi_k^* y \psi_0 \, d\tau$ or $\int \psi_k^* z \psi_0 \, d\tau$. Furthermore, if a particle is randomly oriented with respect to the radiation the average transition probability is proportional to $\frac{1}{3} |\int \psi_k^* x \psi_0 \, d\tau|^2 + \frac{1}{3} |\int \psi_k^* y \psi_0 \, d\tau|^2 + \frac{1}{3} |\int \psi_k^* z \psi_0 \, d\tau|^2$, because the random orientation is equivalent to random polarization of the incident radiation.

A more general example of the importance of transition moments is the proof that electric dipole transitions cannot induce changes of spin state in the absence of interactions such as spin-orbit coupling. Under these conditions ψ_k may be written more fully as $\psi_k(\text{space}).\psi_k(\text{spin})$, and similarly so can ψ_0. The integral in the transition moment is over both space and spin coordinates, although x is of course just a space coordinate. Therefore the transition moment may be rewritten

$$\int \psi_k^*(\text{space}) x \psi_0(\text{space}) \, d\tau_{(\text{space})} \cdot \int \psi_k^*(\text{spin}) . \psi_0(\text{spin}) \, d\tau_{(\text{spin})}$$

The latter integral will be identically zero unless $\psi_k(\text{spin}) = \psi_0(\text{spin})$, i.e. unless the transition does not involve a change of spin state, because different pure spin states are orthogonal. Hence we have derived the selection rule that fluctuating electric fields cannot induce changes of pure spin state. This is of importance in the understanding of phosphorescence.

4.8 Stimulated emission

Let us suppose that we had carried out the entire calculation of section 4.4 for a system starting in state k. We would get exactly the same result, except that k and G would be interchanged everywhere (and ω_k would change sign). This is because $\int \psi_k^* x \psi_0 \, d\tau = \int \psi_0^* x \psi_k \, d\tau$, and because we have a term in $(\omega_k + \omega)$ equal to the one in $(\omega_k - \omega)$. Hence we would have predicted an emission of energy of exactly the same size as the previous absorption. Furthermore, because this emission only differs from the absorption in sign, it would be in phase with and in the same direction as the incident radiation. Such a phenomenon is indeed observed and is known as the stimulated emission of radiation. It is of great importance in low-frequency spectroscopy, and in lasers and masers. We can immediately see that a system in which 50% of atoms or molecules are in state ψ_G and 50% in ψ_k will not be able, taken as a whole, either to absorb or emit any radiation of frequency ω_k. Such a system is said to be 'saturated'. This conclusion can be shown to remain valid even when the perturbation theory is taken, as is proper in this case, to higher order. The more complete theory of such systems reveals several further phenomena, such as coherence in laser emission.

4.9 Other transition mechanisms

(i) Spontaneous emission

A detailed investigation of the interaction of particles with free space can be made to show that particles in excited states can emit radiation spontaneously. This is a mechanism that acts in addition to stimulated emission, and it is of importance only at infrared and higher frequencies. Its classical equivalent is the loss of energy from oscillating electrons in particles due to radiation (rather than energy transfers to other parts of the particle for example). The probability of spontaneous emission in a system may also be deduced by a probability argument (problem 4.7). Spontaneous emission is often referred to as natural emission.

(ii) Relaxation

We have already considered one form of 'relaxation' in section 4.4, where a collision was assumed to shift all the energy levels of a system and then to leave the system effectively back in its ground state. This is quite a likely mechanism for loss of energy, or rather for conversion of electron excitation energy into particle kinetic energy. It is one of the ways that radiation heats substances.

However, other more complex relaxation mechanisms are possible, involving continuous interactions. A particle that interacts with its surroundings will have an extra interaction term, V, in its Hamiltonian. V will contain various electrostatic and magnetic energies, such as the repulsion energy of the electrons of the particle with those of surrounding particles. The potential V will be time-dependent in a complex manner, and (using the ideas of Fourier analysis) will contain some component at each of the transition frequencies corresponding to the separations of energy states. It will therefore cause transitions, in both directions, in an analogous manner to radiation. The interaction $eE_0 x \cos \omega t$ is simply replaced by the particle-particle interaction. The only important difference between this interaction and the interaction with radiation is that in this new case absorption of energy by the particle will be slightly less probable than loss of energy because absorption of energy simultaneously implies loss of energy by the surroundings, which is slightly less likely than gain of energy by them except at infinite temperature. (See sections 6.2 (iii) and 7.3.)

(iii)   Other interactions

The magnetic component, H, of radiation is capable of interacting with the magnetic moment, $\bar{\mu}$, of an electron in order to change the magnetic state of the system. The theory is exactly as before, with the interaction energy being H, $\bar{\mu}$. Magnetic resonance occurs via such transitions, which are called 'magnetic dipole' transitions in contrast to the 'electric dipole' transitions which we have been discussing. 'Electric quadrupole' transitions are also possible, but both these and magnetic dipole transitions are generally much weaker in intensity than electric dipole transitions.

4.10   Scattering

Appendix 1 and problem 4.6 discuss the consequences of extending the theory of this chapter to second-order by feeding the first-order result 4.13 back into the equation 4.12. A new prediction occurs that radiation will be instantaneously absorbed and re-emitted at the excitation frequency. This is observed experimentally as Rayleigh scattering, and because it is an instantaneous process it is in phase with the exciting radiation.[4] An additional scattering is due to normal absorption followed by re-emission (section 4.9(i)) after an unknown but probably very short interval. This latter scattering process, which is only likely near to optical absorptions, will produce the phase lag for scattering that is needed by the classical theory to explain refraction.[5] Thus our quantum theory has

[4] This is dealt with more rigorously in problem 4.5.

[5] A more sophisticated account is given in P. W. Atkins, op. cit., chapter 11.

not lost any of the advantages of our classical theory, except perhaps its relative simplicity.

4.11   Quantized radiation

Our calculations so far have treated the radiation as a classical electro-magnetic wave. Logically, however, we should allow for the fact that radiation of frequency v is also quantized into photons of energy hv. Fortunately this can be shown, by considerations beyond the scope of this book, not to affect the main conclusions that we have already reached, and indeed to provide more rigorous versions of some of the 'proofs' that have been made.

One interesting and important property of photons is that they have an angular momentum of $\sqrt{2}\hbar$, with allowed z-components of $\pm\hbar$, but not of zero. The proof of this is complex, and it is not possible to determine whether the angular momentum is of orbital or of spin type. It means that when a photon is absorbed or emitted by a system then the system must gain or lose the angular momentum of the photon. Several important selection rules may be derived from this. In particular, it means that every vibrational transition of a molecule must be accompanied by some change in the rotational or orbital quantum state of the molecule, corresponding to the necessary angular momentum change.

A beam of photons whose z-component of angular momentum is either $+\hbar$ or $-\hbar$ corresponds in the full quantum theory to a classical wave of right or left circularly polarized light. A wave of right circularly polarized light may be thought of as an ordinary wave to which is added a wave of equal magnitude and frequency, polarized in the y direction and lagging 90° in phase. This is shown in figure 4.3.

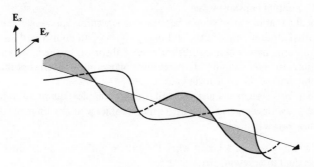

Fig. **4.3** Electric field components of right circularly polarized wave

A brief consideration of this figure shows that the resultant electric field at any one point rotates with the frequency of the radiation, in a clockwise sense looking along the beam in the direction of the wave progression.

The angular momentum of photons can be most clearly demonstrated by irradiating an optically birefringent crystal hanging from a thread with a beam of circularly polarized light along the direction of the thread. The crystal acquires an angular momentum due to the fact that the optical birefringence (different refractive indices for different planes of polarization) destroys the circular polarization of the beam.

Suggestions for further reading
P. W. Atkins, op. cit., chapter 7.
W. Kauzmann, op. cit., chapter 16.

Problems
4.1   Prove equation 4.1 together with the complementary equation

$$+\left(\frac{\partial H}{\partial p_x}\right)_x = \frac{\partial x}{\partial t}$$

What does this suggest about p_x and x?

4.2   How would the conclusion of section 4.4 be affected if E_0 varied significantly over the average radius of the electron orbital? Above what frequencies will such variation occur, and how do these frequencies compare with the minimum frequency of radiation needed to cause ionization?

4.3   Prove without evaluating the integral that

$$\int_{-\infty}^{\infty} \frac{\sin^2(\omega - \omega_k)\tau/2}{(\omega - \omega_k)^2 \tau} \, d(\omega - \omega_k)$$

is independent of τ.

4.4   A hydrogen atom in its ground state is suddenly subjected to a static electric field E_0. Show that the effect of this field may be derived by putting $\omega = 0$ in section 4.4, and that the result is that the coefficients $a_k^* a_k$ fluctuate sinusoidally about a constant value.

   The fluctuations arise simply because one is writing the wavefunction of a distorted atom in terms of the stationary states of an undistorted atom. The constant term, however, measures the energy absorbed by the atom and hence its static polarizability. Express this polarizability in terms of the wavefunctions of the undistorted atom.

4.5   A particle in a system has just two real (i.e. not complex) quantum states, ψ_1 and ψ_2, with energies ε_1 and ε_2, so that its state ψ at any time may be written $\psi = a_1\psi_1 + a_2\psi_2$.

   a Show, using section 4.4, that when the system undergoes a real time-dependent interaction, $p(t)$:

$$i\hbar \frac{da_1}{dt} = a_1\varepsilon_1 + a_2 p_{12} \tag{1}$$

and

$$i\hbar \frac{da_2}{dt} = a_2\varepsilon_2 + a_1 p_{12} \tag{2}$$

where $p_{12} = \int \psi_2 p(t)\psi_1 \, d\tau$, provided that $p_{11} = p_{22} = 0$.

b Show that

$$ i\hbar \frac{d(a_1 a_1^*)}{dt} = (a_1^* a_2 - a_1 a_2^*) p_{12} $$

(You will have to generate an equation for a_1^* by equating the complex conjugates of each side of equation (1))

c Prove that, if no energy is absorbed by the system, $a_1^* a_2 = a_1 a_2^*$, and hence by re-using equations (1) and (2) show that

$$ \psi\psi^* = a_1 a_1^* \psi_1^2 + a_2 a_2^* \psi_2^2 + \frac{2p_{12}(a_1 a_1^* - a_2 a_2^*)}{\varepsilon_1 - \varepsilon_2} \psi_1 \psi_2 $$

This implies that in this steady state the particle density at any point fluctuates in phase or 180° out of phase with p_{12}, about a constant value.

d With a slightly more restrictive definition of a steady state it is possible to generalize the above result for a system with any number of levels. What are the implications of this for scattering theory?

4.6 Show that the first-order equation for a_k in section 4.4 (above equation 4.13) may be approximated to give

$$ a_k \approx e^{-i\varepsilon_k t/\hbar} x_{0k} eE_0 t/i\hbar $$

provided that $\varepsilon_k \gg \varepsilon_k - \varepsilon_0$. Using this approximate expression, and keeping $a_0 = e^{-i\varepsilon_0 t/\hbar}$ as before, solve equation 4.12 to give a better approximation for a_k, and show, by ignoring fluctuating terms in $a_k^* a_k$, that further transitions to state k are likely, with probability proportional to the square of

$$ \sum_{j \neq k, 0} x_{0k} x_{0j} t / (\omega - \omega_k + \omega_j) $$

4.7 A system has two levels, G and E, with populations n_G and n_E respectively. The natural transition probability from the excited state E to the ground state G is A, and the probability of absorption or stimulated emission due to the radiation of intensity u from a black body at the same temperature as the system is uB. Show that, because the system must be in equilibrium with the black body for radiation of all frequencies,

$$ uB(n_G - n_E) = An_E $$

Hence, using equations 1.3, 1.4, and 5.5, prove that

$$ A = 8\pi h\nu^3 B/c^3 $$

Would the equilibrium be possible if B had one value for absorption and another for stimulated emission?

5 Principles underlying spectroscopic techniques

5.1 Electromagnetic spectrum

The actual techniques by which the absorption or emission of radiation by a sample may be detected at different frequencies vary considerably depending on the type of transition to be observed and on the frequency range involved. A detailed description of all available spectroscopic techniques is clearly beyond the scope of this chapter, or indeed, of a whole book. However, this chapter does attempt to show how the choice of technique is largely governed by the frequency of the radiation involved, relative to certain fixed physical constants such as Planck's constant h and Boltzmann's constant k.

The total range of physically available frequencies of radiation is called the electromagnetic spectrum. It is shown in figure 5.1, with the usual names applied to certain frequency ranges and with certain important energies marked at their frequency equivalents. The frequencies are labelled in hertz, but certain other alternative units, such as energy equivalents in $kJ\,mol^{-1}$ and wavelength equivalents in m and nm are included. Another unit which is commonly used in place of frequency is inverse wavelength (cm^{-1}). As the name implies, this is the reciprocal of the wavelength of the radiation measured in centimetres. It is popular with manufacturers of instruments operating in the optical and infrared region. A further unit in common use is the electron volt (eV) which is the potential energy gained by an electron as its electric potential increases by a volt. This unit is popular with some theoretical chemists, and with the designers of particle accelerators. These alternative units are not in themselves units of frequency, but may be related proportionally to frequency with the aid of various fundamental physical constants. The following equalities may be useful:

$$1\,kJ\,mol^{-1} = N_A h \times (2.505 \times 10^3\,GHz)$$

$$1\,cm^{-1} = (29.98\,GHz)/c$$

$$k \times (298\,K) = h \times (6.21 \times 10^3\,GHz)$$

$$1\,eV = h \times (2.418 \times 10^5\,GHz)$$

$$1\,GHz = c/(0.2998\,m)$$

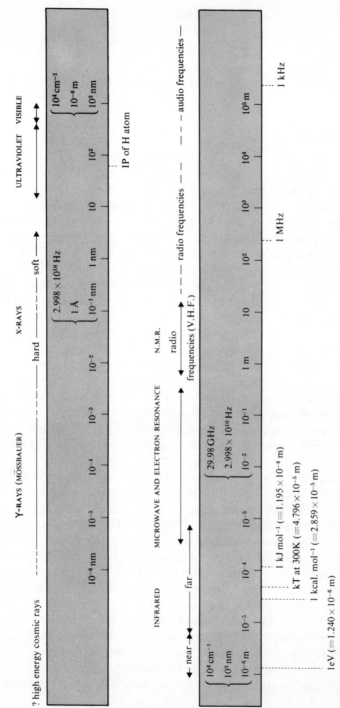

Fig. 5.1 The electromagnetic spectrum

where:

		SI	*cgs*
Avogadro's number,	$N_A =$	$6.0222 \times 10^{23} \, \text{mol}^{-1}$	
Planck's constant,	$h = 6.6262 \times 10^{-34} \, \text{J s}$		$= 6.6262 \times 10^{-27} \, \text{erg s}$
The speed of light in vacuo,	$c = 2.9979 \times 10^8 \, \text{m s}^{-1}$		$= 2.9979 \times 10^{10} \, \text{cm/s}$
Boltzmann's constant,	$k = 1.38062 \times 10^{-23} \, \text{J K}^{-1}$		$= 1.38062 \times 10^{-16} \, \text{erg/}°\text{C}$
Electron charge,	$e = 1.6022 \times 10^{-19} \, \text{C}$		$= 4.8033 \times 10^{-10} \, \text{e.s.u.}$

5.2 γ-ray region

The highest known frequencies of electromagnetic radiation correspond to γ-rays, which are usually produced at certain precise frequencies by the radioactive decomposition of nuclei. The lowest frequency γ-rays may also be identified as X-rays. It is possible to measure the frequency of γ-rays with an accuracy of at least 1% by measuring the size of the voltage pulse produced when a γ-ray photon arrives at a Geiger counter.[1] This pulse height analysis technique is still very much under development, but it is already proving useful in elemental analysis, especially for certain elements such as Co and Fe present in trace quantities. The sample is intensely irradiated in a nuclear pile, after which the unstable nuclei so created are left to decompose, and the resulting γ-radiation is analysed. Each decomposing nucleus produces a characteristic spectrum of γ-radiation, whose intensity is a measure of the amount of the original nucleus that was present. A comparison of the total γ-ray spectrum with the known spectra of samples of known composition leads to an analysis of the unknown sample, at least for those elements which form conveniently unstable nuclei upon irradiation. The actual characteristic γ-ray frequencies are also of great interest in the study of nuclear binding forces because they represent transitions of nuclei between their various bound quantum states.

There is a quite different spectroscopic technique using γ-rays, known as Mössbauer spectroscopy.[2] Briefly, this is a technique for measuring very small shifts in the γ-ray emission and resonant reabsorption frequencies of certain nuclei, such as ^{119}Sn and ^{57}Fe, due to the different chemical or magnetic environments of the emitting and absorbing nuclei. Excited nuclei are initially produced in a standard source by the use of a convenient radioactive parent nucleus (e.g. ^{57}Co for ^{57}Fe). Provided that the emitting and absorbing atoms are in the same environment, and provided this is part of a solid, they emit and absorb at exactly the same frequency. The solid state is necessary in order for the effects of recoil due to photon momentum to be negligible. If they are in different solid states, then the absorber can be brought back into resonant absorption using the Doppler effect (frequency shifts due to motion: see chapter 6). Further shift effects can be observed when the nucleus has a magnetic dipole

[1] R. L. Chase, *Nuclear Pulse Spectrometry*, McGraw-Hill, New York, 1961.
[2] G. K. Wertheim, *The Mössbauer Effect*, Academic Press, New York, 1964.

moment that interacts with a magnetic field, or an electric quadrupole moment that interacts with an electric field gradient at the nucleus. The quantitative explanation of the chemical shifts is still being developed but the technique has already given considerable information about the chemistry of solids containing the appropriate nuclei: for example, about the rôle of the iron atom in haemoglobin and also in nitrogen fixation.

5.3　X-ray region

X-rays are produced in the laboratory by the emission of radiation from moderately heavy atoms (e.g. Cu) which have had electrons from their inner electron shells excited by the impact of high energy electrons, and which are relaxing back to their ground state via a series of electron jumps, the most prominent of which is the final $2p - 1s$ jump ($K\alpha$ line). The frequencies of the jumps are characteristic of the element involved (problem 5.5). Indeed, their original tabulation was invaluable in establishing the concept of atomic number, because the inner shell energies are almost independent of the outer shell wavefunctions, i.e. of chemical environment and periodicity, and depend almost exclusively on atomic number.

It follows from this that if the X-ray emission spectrum of a mixed sample can be obtained, and compared with the spectra of the pure elements, quantitative analysis of the sample is possible. The principle is similar to the γ-ray method of the previous section, except that it is applicable to all elements whose atomic number is sufficient to allow X-ray emission of frequency and intensity high enough for convenient measurement.[3] Magnesium is about the lightest element that is conveniently detectable at present, and the main use of the method is for rapid analysis of alloy samples, for example during steelmaking.

X-ray fluorescence emission[4] is normally generated by irradiation of the sample with a powerful beam of high frequency X-rays, although electron impact is sometimes used to cause emission. The emitted frequencies are analysed using a crystal of known ionic spacing as a grating, and detection is usually by a counter. The detailed frequencies emitted are of some interest to theoretical chemists, especially since small 'chemical shifts' have been observed due to different interactions between inner and outer shells in different compounds of the same atom. The corresponding absorption spectra have been observed using the continuous radiation from a synchotron particle accelerator as a source.

5.4　Ultraviolet and visible region

This spectral region is defined at its lower frequency end by the limits of visibility (ca. 800 nm, 3.75×10^{14} Hz) and at its high frequency end by the

---

[3] See D. J. Fabian (ed.), *Soft X-ray Band Spectra*, Academic Press, London, 1968.

[4] Emission of radiation at lower frequencies following absorption at a higher frequency is known as fluorescence. Nowadays the term is also used to cover all energy loss processes other than re-emission at the original frequency and phosphorescence (ch. 10).

fact that normal prisms and lenses, and air, become opaque above 200 nm, 1.5×10^{15} Hz. Frequencies between this and X-rays can be studied by using gratings, mirrors, and high vacua, but this 'vacuum ultraviolet' region is not of very great importance to molecular spectroscopists in comparison with the ordinary ultraviolet and visible region. This is because the lowest energy absorptions of most molecules occur at lower frequencies than the absorptions of air and quartz, and because these low energy absorptions are the most amenable to theoretical work and hence to identification and correlation.

The ultraviolet and visible region is also a convenient one in that strong continuous frequency light sources are available. Thus, absorption spectroscopy in this region is convenient, which is not true of absorption spectroscopy at higher frequencies. The visible region is easily covered by thermal emission of radiation, e.g. from a tungsten filament, and the ultraviolet region by emission from a high pressure hydrogen lamp, which is continuous in frequency above about 9×10^{14} Hz (330 nm) due to the capture of free electrons with varying amounts of kinetic energy by H_2^+ ions. The capture is a fairly efficient process for relatively low electron kinetic energies, and so the hydrogen lamp is useful in the ultraviolet range but becomes steadily less useful as frequencies increase in the vacuum ultraviolet region.

The light absorbed by the sample may be analysed using a prism or a diffraction grating, the latter being more common in high-resolution instruments. Alternatively, the prism or grating may be used to provide a source of monochromatic light of variable frequency. Detection is by photocells and multiplier tubes, which rely on the photoelectric effect (section 1.5) using metals such as caesium which have very low ionization potentials. A simple spectrometer is illustrated in figure 5.2.

The appearance of an ultraviolet and visible absorption spectrum is dependent on the physical state of the sample. This is unlike X-ray fluorescence spectra which depend almost entirely on inner-shell energies. Outer electron shell energies are considerably affected by intermolecular interactions, and therefore the sample must be in the gas phase for its spectrum to have the properties (such as narrow absorption bands) predicted for isolated molecules and atoms. The very narrow, and hence intense, absorptions of gaseous atoms, e.g. in flames, make their identification particularly easy and this has led to the development of 'atomic absorption spectroscopy' as a method of quantitative element analysis. If the sample is in the liquid phase, the absorption bands will generally be very broad, partly because of the short lifetime of excited states (section 4.5) and partly because the actual energy states of any one molecule will depend to a substantial extent on the molecule's inter-actions at any one time with neighbouring molecules. (Indeed, substantial delocalization of electrons has been shown to occur during collisions in liquids.) In the solid state these interactions will be static rather than dynamic. This means that linewidths in crystalline solids are once again

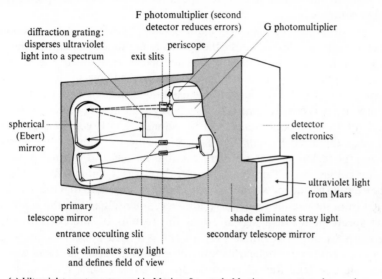

diffraction grating:
disperses ultraviolet
light into a spectrum

F photomultiplier (second
detector reduces errors)

G photomultiplier

periscope

exit slits

spherical
(Ebert)
mirror

detector
electronics

ultraviolet light
from Mars

primary
telescope mirror

entrance occulting slit

slit eliminates stray light
and defines field of view

shade eliminates stray light

secondary telescope mirror

(a) Ultraviolet spectrometer used in Mariner 9 to study Martian upper atmosphere and surface. (The telescope objective replaces the normal laboratory source, sample, and collimator.) Reproduced by permission of NASA, the Jet Propulsion Laboratory of the California Institute of Technology, and the Mariner Mars 1971 Project Office.

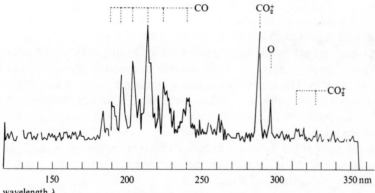

CO CO_2^+

O

CO_2^+

150   200   250   300   350 nm

wavelength λ

(b) Actual spectrum of upper Martian atmosphere, showing ions derived from CO_2, which does not itself absorb in this spectral region. Reproduced by permission of NASA, the Jet Propulsion Laboratory of the California Institute of Technology, and the Mariner Mars 1971 Project Office.

Fig. **5.2**

quite narrow (especially at low temperatures) although the energy levels involved in absorption are those of the whole crystal rather than those of isolated molecules, and hence the interpretation of the spectrum is quite complex.

The other spectroscopic technique using optical frequencies is Raman spectroscopy. In this technique the spectrum of monochromatic radiation scattered from a sample (usually, but not necessary at 90° to the exciting beam) is analysed by conventional techniques. The object is to detect any radiation that is not scattered at the exciting frequency, because the shift in frequency is closely related to the energy separation of the vibrational and rotational quantum states of the sample. Until recently a mercury arc was the usual Raman light source. As the resulting scattering was weak, the technique was not very suitable for gases at normal pressure or for coloured (i.e. absorbing) samples. Nor was it suitable for solids, as the exciting frequency was not strictly monochromatic, so that the reflected light from the sample tended to add a very intense band to the centre of the spectrum whose wings overlaid the regions of interest. However, modern Raman spectrometers use a laser as a light source. As lasers are very powerful and very nearly monochromatic light sources, it has proved possible to extend Raman spectroscopy to cover a wide range of solid, gaseous, and coloured samples.

5.5 Near infrared region

The techniques of optical spectroscopy may be extended to the near infrared region (3.75×10^{14} to 6×10^{12} Hz, 800 to 5000 nm), the limits of which are set by the low frequency infrared absorption continuum of the substance used to contain the sample (and also to some extent of the sample itself). The continuum is due to the wide range of motions attributable to the simultaneous linked oscillations of ions or molecules in the solid or liquid state. The substances with the lowest frequency continua are the alkali halides, especially caesium iodide. These halides are commonly used either as plates between which is supported a thin film of liquid sample, or as discs which are compressed together after mixture of the halide with the solid sample.

The absorption limitations which define the near infrared region also rule out the use of glass and quartz in infrared spectrometers. Instead, one must use mirrors and gratings, as in vacuum ultraviolet spectroscopy. A further limitation that defines the region is that the available continuous radiation sources such as glowing bars and mercury arcs, which all rely on thermal emission, are not sufficiently powerful for conventional absorption spectroscopy below about 5×10^{12} Hz. This may be understood by considering the black body radiation curve in section 1.3. The alternative of studying emission spectra is not generally viable[5] because the

---

[5] Although it is useful in astronomy, where the size of the sample compensates for the low probability of spontaneous emission.

probability of spontaneous emission is proportional to the fourth power of the frequency (see problem 4.7) and is almost negligible at far infrared and lower frequencies.

Infrared radiation cannot be detected by photocells as none of its frequencies are sufficient to eject electrons from any metal. However, a variety of detection methods are available, the basic types being

  i   resistance thermometers or bolometers, in which the radiation is directly absorbed by a resistor and the resulting temperature rise is detected as a resistance change;
 ii   pneumatic detectors, or Golay cells, in which the radiation is used to expand a gas and thus produce distortions in an optical path which can be measured by optical methods;
iii   rapid response thermocouples welded to small, blackened absorbers;
 iv   quantum detectors; these are essentially solid state photocells, in which the radiation ejects electrons not into free space but into the conduction band of semiconductors such as doped germanium, so as to change their conductivity.

At infrared and lower frequencies we meet for the first time the problems of inadequate signal to noise ratios. (Signal to noise ratios are discussed in a general way in the next chapter.) They present a major problem at low frequencies because of the low energy of the quanta involved, and because of the difficulty of distinguishing genuine weak signals due to transmitted low energy quanta from spurious signals due to laboratory disturbances and (at radio frequencies) to the thermal motions of electrons. Spurious signals are largely eliminated in infrared spectroscopy by splitting the incident beam into sample and reference beams and then comparing the intensities of the two after absorption has taken place. This makes the resulting spectrum almost independent of unwanted changes in source or amplification intensity, due to mains voltage fluctuations for example. Variations of this technique are also used in microwave and radio-frequency spectroscopy, and sometimes in optical and ultraviolet spectroscopy also.

Infrared absorption spectroscopy is suitable for samples in all phases, although resolution is poorer with cloudy liquids or mulls and with dispersed solids owing to the fact that not all the light from the sources passes linearly through such samples before reaching the grating, so that collimation is poor. As most samples absorb strongly in the infrared region, pure liquid samples need to be made in the form of thin (0.1 mm) films held between flat halide plates. The great sensitivity of infrared absorption spectroscopy, which is in part due to the narrow widths of most infrared absorption lines, makes it a very useful technique for the study of gases (to obtain molecular constants) and even of unstable molecular species isolated and trapped at considerable dilution in inert solid matrices.

5.6   Far infrared region

The techniques of near infrared spectroscopy are almost useless in the far infrared region because of the weakness of the radiation sources and the difficulties of detection, and also because the long wavelengths are not very suitable for gratings because of interference between different orders of diffraction beam. The sources and detectors would be just about adequate by themselves were it not for the fact that conventional spectroscopic techniques only measure a tiny fraction of the energy output of the source at any one moment in time. As a result of all this, the far infrared region was closed to spectroscopists until recently, and thus little was known experimentally about slow vibrations and coupled vibrational motions. However, the last decade has seen the development of a quite new spectroscopic technique known as Fourier transform spectroscopy[6] which avoids the need for prisms or gratings, is highly suitable for far infrared wavelengths, and uses at any one moment almost all the energy transmitted through the sample in the region to be studied.

In explaining Fourier transform spectroscopy it is not possible to avoid some mathematics because the method actually depends upon the computation of certain integrals. Let us suppose that we wish to study a beam of unpolarized radiation made up of a number of frequencies v_m each contributing a maximum electric field E_m. Let us further suppose, as an initial and reasonable approximation, that v_m/c is confined to integral values, m, of inverse wavelength, i.e. $v_m/c = m$, c being the velocity of light.

The total electric field $E(x, t)$ at a distance x along the wavefront from a point $x = 0$, where all the frequency components have a wavecrest at $t = 0$, will be:

$$E(x, t) = \sum_m E_m \cos \left(\frac{2\pi v_m x}{c} + 2\pi v_m t \right)$$
$$= \sum_m E_m \cos (2\pi m x + 2\pi m c t) \tag{5.1}$$

and

$$E(x, 0) = \sum_m E_m \cos 2\pi m x$$

Clearly the ordinary spectrum of the radiation will simply be the plot of the E_m^2 against v_m. But there is a second way of extracting the E_m; that is by measuring $E(x, 0)$ and then subjecting it to mathematical analysis as follows. Consider the expression

$$\int_{x_0}^{x_0+1} E(x, 0) \cos 2n\pi x \, dx = \sum_m E_m \int_{x_0}^{x_0+1} \cos 2\pi m x \cos 2\pi n x \, dx$$

where x_0 is an integer. Now if $n = m$, then the integral equals $\frac{1}{2}$, except

[6] See P. L. Richards, *J. Optical Soc. America*, **54**, 1474, 1964.

where $n = m = 0$, when it is equal to unity. Also if $n \neq m$ then the integral is always zero. This result has already been used in discussing the ortho-normality of states with different magnetic quantum numbers, and is easily proved. We therefore have

$$\int_{x_0}^{x_0+1} E(x, 0) \cos 2m\pi x \, dx = \tfrac{1}{2}E_m (m > 0) \quad \text{or} \quad = E_0 (m = 0) \qquad (5.2)$$

and therefore

$$\int_0^L E(x, 0) \cos 2m\pi x \, dx = \tfrac{1}{2}LE_m \, (m > 0) \quad \text{or} \quad = LE_0 \, (m = 0) \qquad (5.3)$$

provided that L is a large number. This is because L may be written as an integer L_0 plus a fractional part l. There will be L_0 integrals as in equation 5.2 within the range $0 - L_0$, and the extra contribution to the integral in equation 5.3 from the range L_0 to $L_0 + l$ will be negligible (or, more precisely, will represent the small irreducible error of this system of measurement).

With the aid of equation 5.3 we can thus calculate E_0 and all the E_m by performing a fairly straightforward integration, provided that we have our raw data $E(x, 0)$ as a function of x. We extract $E(x, 0)$ experimentally by splitting the beam represented by $E(x, t)$ (equation 5.1) into two equal halves, and then recombining the halves after one has passed through a distance X and the other through a distance $X + 2\delta$. This is easily done using an interferometer. The amplitude at the point of re-combination will be

$$E(X, t) = \tfrac{1}{2} \sum E_m \cos (2\pi m X + 2\pi m c t)$$
$$+ \tfrac{1}{2} \sum E_m \cos [2\pi m (X + 2\delta) + 2\pi m c t] \qquad (5.4)$$

Now $\cos A + \cos B = 2 \cos \tfrac{1}{2}(A + B) \cos \tfrac{1}{2}(A - B)$ and therefore:

$$E(X, t) = \sum_m E_m \cos [2\pi m (X + \delta) + 2\pi m c t] \cos 2\pi m \delta$$

Hence the r.m.s. electric field amplitude in the beam at any point will be

$$\tfrac{1}{2} \sum_m E_m \cos 2\pi m \delta = \tfrac{1}{2}E(\delta, 0)$$

Thus, if we measure the intensity of the recombined beam as a function of δ, in the range 0 to L, and then write x in place of δ, we obtain $E(x, 0)$ and hence our various E_m and hence our normal spectrum. The final stages of integration and plotting of E_m^2 against v_m are easily and auto-matically performed by a computer.

The above mathematics assumes the validity of Fourier expansions. To deal with a continuous range of frequencies it is necessary to use the full theory of Fourier transforms, which is an extension of Fourier ex-pansion theory. This explains the name of the spectroscopic method.

5.7 Microwave region

Even far infrared spectroscopy cannot be made to extend much below 7.5×10^{11} Hz (i.e. to wavelengths greater than 0.4 mm). It is necessary instead to use electronic oscillations to produce the radiation. We have already seen how oscillating electric charges emit radiation through space. Fortunately, frequencies up to nearly 10^{12} Hz are obtainable using both vacuum and solid state oscillators, the former being known as klystrons. However these oscillators are not as yet fully tunable over wide ranges of frequency. Also not much power is available at the higher klystron frequencies. Hence certain ranges of the microwave region are at present more available for absorption spectroscopy than others, although the situation is rapidly improving with the development of backward wave oscillators. Fortunately the molecular rotations studied by this method each produce many absorptions providing partly redundant information over quite a wide frequency range. The reasons for this will be discussed later. Several absorptions due to any one molecule are likely to be detectable even though it is not possible to study absorptions at all frequencies, and hence much molecular information can be extracted.

The advantages of using tunable monochromatic radiation are, of course, that no mirrors or gratings are necessary, and also that quite high radiation intensities can be achieved. Detection of these relatively high intensities is easy, using the same thermal devices as in infrared spectroscopy, and all that remains is to collimate the radiation into a beam that traverses the sample. This is readily done within limited frequency ranges by confining the radiation within internally polished metal pipes of rectangular cross-section, known as waveguides. The most efficient transmission occurs when the long side of the cross-section rectangle equals the wavelength, which is another reason for the high frequency limit of microwave spectroscopy.

Simple absorption spectroscopy may be carried out on gaseous samples merely by filling the entire waveguide with the gas or vapour to be studied. Normally, such absorption spectroscopy is only of interest with gaseous samples, because the quantum jumps one wishes to study are those between rotational levels, which are either non-existent or non-observable in the vast majority of liquids and solids. It is also easy with this method to study the effect of subjecting the gas to an electric field (the Stark effect). This is done by inserting a thin insulated metal strip down the centre of the entire length of the waveguide, parallel to the wider face, and then applying a voltage to the strip. Figure 5.3 gives an outline diagram of a simple microwave spectrometer.

In electron resonance spectroscopy similar methods are used to study the rather weak microwave absorptions of free radicals in a strong homogeneous magnetic field. As the sample and the homogeneous part of the magnetic field are usually both quite small and the absorption intrinsically weak, it is necessary to pass the microwave beam through the sample many times. This is done using a device known as a microwave cavity, which is a hollow metal-lined volume whose dimensions are closely related to the wavelength of the microwave radiation in order to produce standing waves and hence minimum absorption at the walls. The optical equivalent of a microwave cavity would be a tiny box lined with highly efficient mirrors (except for a small entrance hole). The hole in the box, viewed from outside, would itself appear to be a perfect mirror unless absorption occurred inside the box. Hence a comparison of the intensities of the light going into and out of the hole would be a sensitive test of absorption within the box and would, furthermore, be independent of small unintended fluctuations in incident light intensity, as with double

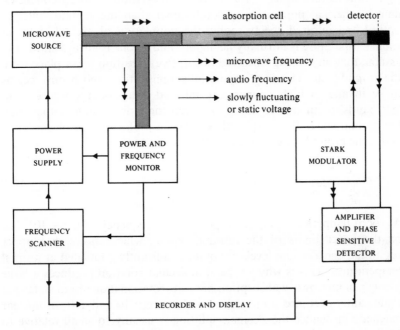

Fig. **5.3** A simplified single-cell microwave spectrometer

beam infrared spectroscopy. A typical microwave cavity produces the equivalent of about 5000 passes of the microwaves through the sample and is said to have a quality factor, or Q, of 5000. Furthermore, by varying the position of the sample within the cavity it is possible to select regions where only the magnetic or the electric field components of the radiation are individually large. This is useful when both electric dipole and magnetic dipole absorptions occur simultaneously. Because of the unusually strong electric dipole absorption by water at all frequencies[7] it is quite difficult to carry out electron resonance experiments on aqueous samples unless they are very carefully confined to the region of the cavity where the radiation electric field is zero; the electric dipole absorptions

[7] The very high microwave absorption of water is used to advantage in microwave ovens, which heat all but the driest foods rapidly, leaving the container almost cold.

otherwise 'spoil' the cavity Q for the study of magnetic dipole absorptions. However, apart from this, the method is suitable for studies of samples in all phases.

Further sensitivity is attainable in microwave studies, as it also is in infrared and radio-frequency spectroscopy, by the device of lock-in detection which is discussed in the next chapter.

5.8 Thermal population and stimulated emission

Microwave and magnetic resonance absorption spectroscopy share a significant feature which is relatively unimportant at higher frequencies, namely, thermal population of levels other than the ground state. In optical and infrared spectroscopy, the population of excited states at room temperature is normally negligible, and so one is always entitled to assume that any strong observed absorptive transition takes place from the ground state. This is a very useful simplification, and it only begins to be invalid in high temperature infrared spectroscopy, where extra 'hot bands' occur due to absorptive transitions from states other than the ground state. The fractional population p of a state n of energy ε_n above the ground state is given by Boltzmann's law as

$$p = \frac{e^{-\varepsilon_n/kT}}{\sum_n e^{-\varepsilon_n/kT}} \tag{5.5}$$

At room temperature, kT corresponds to a frequency of 6.2×10^{12} Hz, on the lower limits of the infrared region, which shows that upper rotational and Zeeman levels are quite significantly populated at normal temperatures. This is why a typical molecular rotation produces a large number of microwave absorption lines: its rotational quantum states are significantly occupied up to values of J between 10 and 100. In nuclear magnetic resonance the Zeeman splitting is usually so small relative to kT that the various nuclear orientational states differ in population by less than one part per million. (This is a very inconvenient fact for nuclear physicists who wish to orient nuclei for scattering experiments, as they are forced to use very low temperature devices and also additional techniques in order to achieve even partially aligned nuclear targets.)

The population of upper states has a second disadvantage which is more serious than that of merely complicating the spectrum. We saw in the last chapter that the mechanism which permits absorption from the ground state to an excited state also permits stimulated emission from the excited state to the ground state, to an identical extent. It follows that, if a sample with n excited states contains a mixture of molecules with exactly $100/(n+1)\%$ in the ground state and $100/(n+1)\%$ in each excited state, then no net energy absorption or emission can occur when radiation interacts with the system, for absorption and stimulated emission will always balance exactly. Therefore absorption of radiation at low frequencies depends on the population *difference* between the levels under study. Hence, the lower the absorption frequency, the weaker the absorption, quite apart from the low energy of the quanta involved. It is partly for this reason that much money is spent in nuclear magnetic resonance spectroscopy in order to achieve very high magnetic fields.

The population difference between levels predicted by equation 5.5 depends upon the establishment of thermal equilibrium, which in turn depends on the energy jumps of the resonating nuclear or electron system induced by interaction with its surroundings. These energy jumps also usually determine the lifetime of the various states and hence the line-widths of the absorptions which link them. Thus the low-frequency spectroscopist always faces a dilemma. If thermal equilibrium is rapidly established in the system he is studying, his spectral linewidths will be large and so he will lose both resolution and sensitivity. However, if thermal equilibrium is only slowly established, then he can only use low powers of radiation, for radiation tends to equalize the populations of the states it links when the transitions it causes become more probable than natural thermal transitions (see section 7.3(ii)). Thus, he has to confine his attention to systems with convenient rates of establishment of thermal equilibrium (relaxation times) and even with such systems he has to set his radiation intensity at a level high enough to allow accurate detection but low enough to avoid equalizing energy level populations (saturation). This provides serious limits to the sensitivity and applicability of microwave, electron resonance and nuclear magnetic resonance spectroscopy.

5.9   Radio-frequency spectroscopy

Radio-frequency spectroscopy in its simple, continuous irradiation form is used to study the transitions of nuclei either between the orientation states of their magnetic dipoles in a strong external magnetic field, or between the orientation states of their nuclear quadrupoles in the internal electric fields of the molecule within which they are contained, or between the orientational states in a combination of these. The frequencies involved are usually between 10^6 and 10^8 Hz, i.e. between medium wave radio frequencies and VHF. Therefore, for the reasons outlined in the previous section, the methods have inherently low sensitivity, and require rather sophisticated apparatus. Even if a sufficiently large magnetic field is obtainable, the radio-frequency method becomes progressively less sensitive than expected above 10^8 Hz because conductors carrying alternating current above this frequency radiate strongly. This makes circuit design much more difficult than at lower frequencies.

The simplest part of a radio-frequency spectrometer is the source. Continuously tunable oscillators are readily available in the megahertz frequency range, capable of providing as much power as is normally required. The detection of absorption is more difficult. Clearly the use of a microwave-type cavity is impossible with such long wavelengths of radiation. Instead, the sample is usually contained within a tuned coil or equivalent device. The oscillating magnetic field produced in this coil by the a.c. current in the coil, provided it oscillates at the resonance frequency, sets the nuclei into coherent oscillations. These produce their own oscillating magnetic field, which can be detected either as a detuning of the original coil or as a signal in a coil whose axis is at right angles both to the original coil and to the field direction.

In more detail, the process is as follows. The nuclei precess about the main magnetic (or electric) field direction, which we define as the z-direction, with a characteristic circular frequency which we wish to discover. In the absence of any radio-frequency fields the precessions are of random phase, and there is thus a net magnetic moment pointing in the z-direction proportional to the population difference between the energy levels. Now let us suppose that a radio-frequency magnetic field at or near the precession frequency is applied in the y-direction using a coil. At this point it is useful to remember that any linear oscillation can always be considered as two superimposed circular oscillations with opposite rotation directions. Thus, in addition to the main field, the nuclei will experience a weaker field whose direction rotates about the z-axis at or very near their circular precession frequency. The net magnetic moment will therefore undergo forced, damped oscillations more or less as described in chapter 4, except that the motion will be circular and the amplitude will be measured by the deviation of the magnetic moment from the z-axis. The damping may be identified with relaxation, i.e. the rather infrequent natural changes of direction of the individual nuclear spins due to mutual and lattice interactions. In other words, the magnetic moment will go into forced precession at the driving frequency, although not necessarily in phase with it. This rotating net nuclear magnetic moment will itself produce a current in the original coil, which at the resonance frequency will be 90° out of phase with the original current. The phase lag will be detectable electronically, and will (as in double beam experiments at higher frequencies) be independent of source power fluctuations, which is important with such weak absorptions. Alternatively the field of the rotating nuclear moment can be detected by a coil whose axis is in the x-direction and therefore which is unaffected by the driving frequency. This is another way of obviating the effect of source power fluctuations, and once again one uses electronic means to detect the phase lag of the nuclear rotation relative to the driving field. The phase lag is measured because the absorption line shape is the same as the variation of the out-of-phase oscillation amplitude with frequency as was explained in chapter 4.

The next problem in radio-frequency detection, as in all radio reception, is to distinguish the electron oscillations in the receiver coil caused by the nuclear motion (the signal) from the random thermal motion of the electrons in the coil (the noise). Needless to say, it is very rarely possible to cool the coil to within a few degrees of absolute zero, although this would certainly reduce noise drastically. One therefore has to resort to devices such as lock-in detection and relatively slow scanning using frequency filters. These methods are discussed in chapter 6, as they are relevant to several areas of spectroscopy. A block diagram of an n.m.r. spectrometer is given in figure 5.4.

With nuclear magnetic resonance we have already seen the importance of using the highest obtainable external magnetic field. A partially conflicting requirement is that the field should be highly homogeneous over the width of the sample. The natural linewidth of nuclear resonance absorptions is often less than 1 Hz, and one often wishes to resolve absorptions with such a frequency separation. It is not difficult to create radio-frequency oscillations which are stably monochromatic to within 1 Hz, even though this implies a stability of one part in 10^8 at a total frequency of 10^8 Hz. But because the absorption frequency also depends upon the external field (see chapter 7) it is also necessary for this field to be stable in time, and homogeneous over the sample volume, to within this very severe tolerance. It is worth remembering that one part in 10^8 is equivalent to a distance of 0.4 m relative to the circumference of the earth! Such stability and homogeneity can be provided, at a price, by using large, accurately machined magnets plus a variety of devices which compensate for field fluctuations, link field to frequency, and correct for remaining inhomogeneity by using low current supplementary solenoid coils ('shim coils'). Homogeneity can also be considerably improved by spinning the sample about the y-axis (see chapter 7).

Because electron and nuclear Zeeman level separations are proportional to the external field, at least within certain ranges, magnetic resonance spectroscopists have the option of obtaining their spectra by varying the external magnetic field rather than the frequency. This is often convenient, especially in electron resonance, when the source, circuitry and cavity only work well within a limited frequency range.

5.10   Fourier transform pulse spectroscopy

Conventional nuclear magnetic resonance spectroscopy relies on a slow scanning of the resonance region so that only one frequency (or field) is used at any one moment. This is experimentally simple, but with complex, spread-out spectra it has the disadvantage of taking a long time. It is also not very suitable for resonances (such as those of ^{13}C) which are easily saturated.

The alternative technique of Fourier transform pulse spectroscopy[8] is currently under development as a means of avoiding these difficulties. The principle of the technique is closely related to the far infrared method described in section 5.6. The sample is subjected to very brief bursts (pulses)

[8] R. D. Becker and T. C. Farrar, *Pulse and Fourier Transform N.M.R.*, Academic Press, New York, 1971.

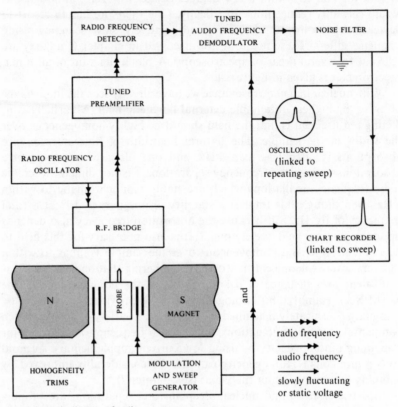

(a) Schematic diagram of ordinary n.m.r. spectrometer

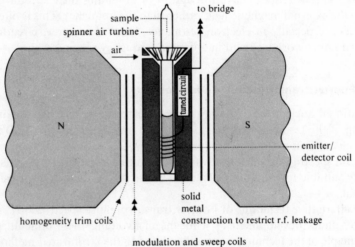

(b) Diagram of probe region (horizontally expanded)

Fig. **5.4**

Table 5.1 A summary of spectroscopic methods

Approximate frequency range available	Method	What is detected	Special features	Applicability	Main uses
above 10^{18} Hz	γ-ray emission	emitted γ-rays from nuclear transitions	specific for certain nuclei	condensed phases	quantitative analysis
" "	Mössbauer	resonance reabsorption of γ-rays	specific for certain nuclei	solids only	chemical bonding studies
$10^{17} - 10^{19}$ Hz	X-ray fluorescence	emitted X-rays from inner shell	almost independent of chemical state	condensed phases	quantitative analysis
1.5×10^{15} Hz upwards	Vacuum ultraviolet spectroscopy	high energy electronic transitions	mostly emission measured	particularly useful for vapour phase	data for theoretical calculations
4×10^{14} to 1.5×10^{15} Hz	Near u.v. and optical spectroscopy	low energy electronic transitions	mostly absorption measured	all phases	identification, analysis and theory; kinetic studies
"	Raman spectroscopy	vibrational transitions	relies on scattered light	all phases	identification and analysis
"		also a very few rotational transitions	—		
6×10^{12} to 4×10^{14} Hz	Near infrared				
7×10^{11} to 6×10^{12} Hz	Far infrared	rotational transitions	Fourier transform spectrometer needs a computer		
3×10^{9} to 7×10^{11} Hz	Microwave	rotational transitions	molecules must have dipole moment	vapour phase, simple molecules if structural data required	molecular structure measurement
Approximate frequency range available	Method	What is detected	Special features	Applicability	Main uses
3×10^{9} to 10^{11} Hz	Electron resonance	Zeeman transitions of unpaired electrons	requires strong magnetic field	free radicals, all phases	identification of free radicals, and free electron distribution studies
10^{6} to 3×10^{8} Hz	Nuclear magnetic resonance (including pulse transform methods)	nuclear Zeeman transitions	requires very strong, highly homogeneous magnetic field	all phases, but mainly to liquids	identification and kinetic studies
10^{6} to 2×10^{9} Hz	Nuclear quadrupole resonance	quadrupole Stark effect in internal field	limited to nuclei with $I > 1$. Not very sensitive	solids only	electron distribution in bonds

of a radio-frequency field of high power. Provided this power is sufficiently high it will stimulate the rotation of the net magnetic moment of all the nuclei within a frequency range wide enough to cover the entire spectral region of interest. When the pulse ceases, the net moment of each chemically distinct group of nuclei will precess at its own natural (Larmor) frequency, and this precession will be detectable as a time-dependent voltage $V(t)$ induced in the receiver coil of the spectrometer. $V(t)$ may be compared with $E(x, 0)$ in equation 5.1, and is indeed almost the same as $E(0, t)$, the radiation field at the source measured as a function of time. It follows that

$$\int_0^T V(t) \cos 2\pi mt \, dt = \frac{T}{2} V_m \, (m > 0) \quad \text{or} \quad T V_0 \, (m = 0)$$

where $V(t)$ is assumed, for simplicity, to contain only the frequencies v_m, where $v_m = m$, an integer. 0 to T is normally taken as the time interval between pulses, after which the measurement is repeated to improve accuracy. Hence, with the aid of a computer, we can extract the various precession frequencies and peak intensities that we want from $V(t)$. No interferometry is needed because it is possible to measure $V(t)$ directly at radio frequencies. The method is easiest to analyse when relaxation is ignored, so that $V(t)$ is a non-decaying voltage made up of one frequency component per resonance line. But it is equally valid when $V(t)$ does decay. In this case the calculated spectrum appears with finite linewidths. Furthermore, the rate of decay of $V(t)$ may be measured directly for each component, and thus relaxation times may be extracted rather accurately.

Pulse techniques may also be used to extract magnetic resonance information without the use of Fourier transform computations. However, the theory of the coherent spin motions that are involved is beyond the scope of this book.

There is in principle no lower frequency limit to magnetic resonance spectroscopy. The only limitation is that of ever decreasing signal to noise ratio due to the tiny population differences between upper and lower states. No other form of low frequency spectroscopy exists at present. All the spectroscopic methods mentioned in this chapter are summarized in table 5.1.

Suggestions for further reading
As references in text, plus
D. H. Whiffen, *Spectroscopy*, Longmans Green, London, 1966.
S. Walker and H. Straw, *Spectroscopy*, Chapman and Hall, London, 1961, vols. I and II.
W. J. Moore, *Physical Chemistry*, Longmans Green, London, 1956 et seq., chapter 14.
R. A. Sawyer, *Experimental Spectroscopy*, Dover, New York, 1963.

Problems
   5.1   Calculate the following quantities, indicating the fundamental constants that must be included to make the relationships dimensionally exact:

    a the ionization potential of the hydrogen atom, 13.4 eV, in Hz and cm^{-1};
    b the wavelength of radiation of frequency 60 MHz;
    c kT at 25 °C in cm^{-1};
    d the frequency of radiation of wavelength 0.1 nm in Hz and eV.

5.2 Prove from first principles that $1\,eV = 2.418 \times 10^{14}\,Hz \times h$. What is the maximum frequency of radiation that can be produced by electrons that have been accelerated through 50 kV and then allowed to strike a heavy metal target? What is the wavelength of this radiation?

5.3 What spectroscopic methods could you use to investigate:
    a the percentage of alcohol in a beverage;
    b the yellow colour produced (reversibly) by heating zinc oxide strongly in air;
    c whether the mercurous ion in solution is Hg^+ or Hg_2^{2+};
    d whether an enzyme contains small amounts of cupric ion?
    In all these examples you should consider more than one method.

5.4 Explain carefully why none of the spectroscopic methods described in chapter 5 are of much use for following a chemical reaction taking place in a liquid metal solvent.

5.5 An approximate formula for the frequency v of the $K\alpha$ line of an element of atomic number Z is $v^{1/2} = a(Z - b)$ where a and b are constants, a being near to $\sqrt{3R_\infty}/4$ and b being between 0 and 1. Explain why this might be so.
    A better fit with experiment is gained if a and b are given separate values for each row of the periodic table. Why?

5.6 Explain why yellow or brown is a much commoner colour for a natural material than is blue. Use your answer to explain why most natural colours are richer in countries with little sun than in countries where the sun shines regularly. Also explain why many substances absorb light by electronic transitions at optically visible frequencies, but few do so by electronic transitions occurring below visible frequencies.

5.7 Using equation 1.3, calculate the relative intensities of radiation from black bodies at 1500 K and 2000 K, at the two limits of the near infrared region (4.75×10^{14} and 6×10^{12} Hz).

5.8 A pulse of radiation of frequency v is emitted for a time $10/v$ seconds. Use the arguments of sections 5.6 and 5.10 to show that the uncertainty in frequency at a detector is of the order of 10%.
    What factors do you think determine the choice of the time interval T between pulses in Fourier transform pulse n.m.r.?

5.9 Calculate the relative populations at 4 K and 300 K of states that would be linked by radiation of frequencies of 10^8, 10^{10}, and 10^{12} Hz.

5.10 What magnetic field would be necessary to produce 50% alignment of protons (i.e. a 2:1 Zeeman population ratio) at a temperature of 1 K? (Note: some of the information needed to answer this question is in chapter 7.)

5.11 Using the result of problem 4.7, calculate the frequency at which $A = B$, in Hz and cm^{-1}.

5.12 Why is the Fourier transform pulse method of n.m.r. more appropriate for complex spectra than for simple ones?

5.13 Suggest experimental ideas that would:
    a reduce the radiation reaching a light-sensitive sample to the minimum necessary to record an ultraviolet absorption spectrum;

b reduce to a minimum the scattering from an insoluble solid sample in the beam of an infrared spectrometer;

c permit the observation of free radicals with lifetimes of ca. 1 second produced by the reaction of two different solutions in a non-polar solvent;

d enable one to obtain almost independent spectra of two superimposed n.m.r. resonance lines of very different width but of equal area.

6 Quality of a spectrum

6.1 Resolution and resolving power

The existence of energy absorption or emission by a system does not guarantee that a spectrum will actually be observed, for the detectable effect, which is known in general terms as the signal, may be too weak for reliable recording. This is particularly likely if the linewidth of the signal is large, because then the peak height for a given energy absorption or emission rate will be correspondingly small (section 4.6).

Furthermore, even if a signal is recorded, it will be relatively useless if it overlaps with many other signals so that the individual quantum jumps causing the signals cannot be identified. There are two ways in which one may attempt to reduce or avoid this overlap and thus improve the resolution of the spectrum. The first is to reduce the linewidth of the absorption and the second is to increase the resolving power of the instrument. Linewidths are commonly measured in Hz and reciprocal resolving power in parts per million. Together they determine the resolution, according to the relation:

$$\text{linewidth} + \frac{\text{total frequency}}{\text{resolving power}} = \text{resolution}$$

The resolving power of a spectrometer is the measure of its ability to separate different frequencies of radiation, either before absorption by the sample or after emission or absorption. Figure 6.1 shows two narrow absorption lines under various degrees of instrument resolving power. The situation in (c), where a dip is just distinguishable between the peaks, is normally taken as the point where the two lines are resolved; at this point the resolving power is defined as the total frequency divided by the frequency separation of the underlying absorption lines. The line shape and width of the individual peaks in (e) depends on how the instrument separates different frequencies, provided that the true linewidth is considerably less than the peak linewidth.

The resolution of a noisy spectrum may be improved somewhat in a practical sense by using an amplifying and recording device which draws out the first or second derivative of the peak height with respect to the frequency. This is discussed in section 6.3.

The lack of resolving power is a serious limitation on the resolution of γ-ray and X-ray spectrometers. This is because total emission and absorption is relatively weak, especially with samples containing only

peak heights not to scale

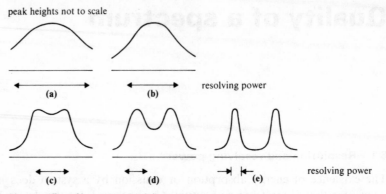

Fig. **6.1** Effect of instrument resolving power

small amounts of the elements of interest. It follows that the more this radiation is split into its component frequencies, the less is energy transmitted at any one frequency. Eventually the rate of arrival of photons within the chosen frequency range that is separated out (the bandwidth) is so small that substantial statistical fluctuations occur in the count rate unless the count takes place over a very long time. The word 'noise' is used to describe such statistical fluctuations, and also any other fluctuations in the recorded spectrum, such as those caused by background radiation or by faulty circuitry, which are not attributable to the signal. The statistical fluctuations are an irreducible source of noise in that, for a given sample, spectrometer, and frequency of observation,

$$\frac{\text{resolving power} \times \text{maximum signal intensity}}{\text{r.m.s. noise intensity} \times \left(\begin{array}{c}\text{time of observation}\\\text{of fixed frequency range}\end{array}\right)^{1/2}} = \text{constant} \quad (6.1)$$

The appearance of the square root is explained via problem 6.2. It follows from expression 6.1 that to double the resolving power requires four times as long a scan for a given sensitivity; for this reason, the resolving power in spectrometers at optical and higher frequencies is normally set no higher than is necessary in order to extract the required information from the spectrum that one obtains.

The constant in expression 6.1 depends on the available source power at the frequency concerned, at high frequencies. Although source power is usually rather low for γ-ray and X-ray spectroscopy, it is usually sufficiently high in optical and near infrared spectroscopy to mean that the resolving power of the spectrometer is considerably greater than the linewidth of the signals, particularly those from liquids and solids. When this is so it is possible to arrange for the resolving power of the spectrometer to adjust automatically within limits by variation of the detector slitwidth, so that a constant voltage is recorded at all frequencies in the absence of absorption. This permits the rapid comparison of absorption intensities at different frequencies without the need for detectors whose response is linear over a very large energy range. It also permits the standardization of 'fingerprint' spectra (section 9.6).

In Fourier transform spectroscopy the instrument resolving power is again dependent on the time of scan, because for a given sensitivity this time determines the total range of the integral that is calculated, and hence the extent to which it approximates to the ideal case of an integration from zero to infinity. It also determines the number of points measured for the summation by which that integral is in fact approximately calculated.

In magnetic resonance spectroscopy it is commonly (but not universally) true that the resolving power of the spectrometer is determined by the inhomogeneity of the magnetic field over the volume of the sample. In this case expression 6.1 only holds in the sense that the resolving power is fixed for a given instrument prepared in a given way. In fact if the instrument can be adjusted to improve the resolving power, the maximum signal intensity will increase in proportion because the area under the spectral line is proportional to the energy absorbed by the sample.

6.2   Linewidths

The linewidth of a spectral absorption is normally defined as the total width at half height. An understanding of what determines the width of spectral lines is valuable to the spectroscopist, both because it may help him to improve the resolution of a spectrum and also because it may give him an insight into the static and dynamic interactions that are experienced by the atoms or molecules he is studying. He may even be able to distinguish between different possible interactions by an accurate observation of the spectral line shapes. The main factors that determine linewidths are listed below.

(i)   Static interactions

The energy states of atoms and molecules in condensed phases may be quite strongly dependent on their surroundings at any one moment in time. The resulting frequency spread in a spectral absorption of the system may be the main factor in determining the linewidth and shape of the absorption. This was mentioned in section 5.4 as one explanation of the width of optical absorption lines in liquids. The line shapes caused by static interactions may be thought of as the sums of a large or infinite number of sharper absorption lines. These lines may each be due to a small number of molecules in a specific state, or to vibrational or rotational fine structure, or to one complex vibrational motion of a molecule in a liquid or crystal which approximates, along with motions of similar frequency, to a motion of the isolated molecule. The linewidth in this last case is a measure of the interaction of the molecule with its surroundings; broad infrared absorption lines or broad vibrational fine structure lines in optical spectra are associated with strong interactions such as hydrogen bonding. A final example of static broadening is that of a magnetic resonance line whose resolution is limited by the magnetic field inhomogeneity.

(ii) Doppler effect

The change in pitch of the alarm bell as a fire engine rushes past is a well-known example of the Doppler effect of motion upon sound waves. It occurs equally when the observer moves rapidly past a static source of sound, and arises because the speed of sound in air is a constant under given conditions. The speed of light in vacuo is similarly constant, and hence the precise frequencies of radiation which are absorbed or emitted by atoms or molecules depend on their component of velocity in the direction of the radiation at the time of interaction. The likely spread of these components is easily calculated for a gas from equation 5.5, where ε_n is interpreted as the kinetic energy $\frac{1}{2}mv_r^2$ due to motion in the direction of the radiation. The probability that the component is v_r is simply proportional to $e^{-mv_r^2/2kT}$. This has a maximum at $v_r = 0$ and half that value at $mv_r^2 = 2kT \ln 2$. A velocity component v_r produces a fractional shift of frequency of v_r/c, and hence the resulting linewidth l, in frequency units, of a line at frequency v due to Doppler motion is

$$l = \frac{2v}{c}\left(\frac{2kT\ln 2}{m}\right)^{1/2} = \frac{2.235v}{c}\left(\frac{kT}{m}\right)^{1/2} \tag{6.2}$$

which for a typical molecule at room temperature means a broadening of about one part per million. If the Doppler effect is the dominant cause of broadening, the spectral line will have a Gaussian ($y = e^{-x^2}$) shape, which in comparison with the Lorentzian shape (figure 4.2) has a broader peak and smaller wings for a given linewidth and height.

The Doppler effect is used in Mössbauer spectroscopy to cancel the difference in static absorption and emission frequency between source and sample. As the frequencies involved are of the order of 10^{19} Hz and the shifts of the order of 10^7 Hz, it is clear that provided the sample or emitter can be given a range of velocities up to $\pm 1\,\mathrm{cm\,s^{-1}}$ all likely Mössbauer shifts can be easily covered.

The effect is also used to explain the frequency shifts in the radiation from galaxies and stars, and thus to calculate their velocities. However, it is not the only possible interpretation of the shifts.

(iii) Continuous dynamic interactions

A static interaction becomes a dynamic one, in spectroscopic terms, when it fluctuates at a higher frequency than the frequency equivalent of the line shift it produces. The difference between the two cases is explained by reference to an example in section 7.4 (ii). More formally, one may follow through the perturbation argument in chapter 4 with a second time-dependent interaction in the Hamiltonian. If the interaction fluctuates slowly it will be almost constant within the time taken to absorb or emit photons, and will simply shift the various energy levels at any one moment by a fixed amount. This is the static case. If it fluctuates very rapidly, as is usually the case in low frequency spectroscopy, then its motion may be Fourier analysed into separate frequency components, one of which will be at the absorption or emission frequency and one at zero frequency. The zero frequency component will usually be small, but if it is not, it will contribute to a line shift. The size (in energy units) of the resonance frequency component will determine the linewidth. This is an important fact which can only be derived accurately from quite complex relaxation theory. However, it may be approximately understood by considering the continuous relaxation interaction as a means of discontinuous exchange of energy quanta between the system and its surroundings. The probability of loss of an energy quantum will be proportional to the energy of the resonance frequency component of the relaxation interaction, just as the probability of a radiation induced transition is proportional to the radiation energy, and the loss will correspond to the collision process that was described in section 4.4 as an explanation of linewidths. Calculations on this and related models show that the half linewidth of the absorption line measured in frequency units is equal or close to the reciprocal of the relaxation time, this being the average lifetime in the excited state (problem 6.5).

The intensity of the lattice motion component of frequency ω is known as the spectral density, $J(\omega)$. It is possible, using the statistical theory of random fluctuations, to calculate $J(\omega)$ as a function of ω for random lattice motions. The result is

$$J(\omega) \propto 1/(1+\omega^2\tau_c^2)$$

where τ_c is called the 'correlation time' of the lattice motions, and is approximately the average time taken by a lattice motion, e.g. between collisions. The above formula may be understood better by reference to a highly simplified model. Let us consider a lattice which consists only of rotating classical electric or magnetic dipoles whose moment of inertia is I and whose various rotation frequencies are ω. Most of the rotations will produce (at a distance) electric or magnetic fields fluctuating with frequency ω. The probability that a rotor will have kinetic energy $\frac{1}{2}I\omega^2$ is proportional to $e^{-I\omega^2/2kT}$, according to Boltzmann's law, and therefore $J(\omega)$ is approximately proportional to

$$e^{-I\omega^2/2kT} \approx 1/(1+\omega^2\tau_c^2)$$

if ω is not too large and if $\tau_c \approx \sqrt{I/2kT}$. This latter time is slightly less than a typical rotation time, and thus comparable to the average time between collisions.

If T is lowered, τ_c will lengthen, and will become considerable at the point where free rotation no longer occurs. Therefore, lowering the temperature of a sample will cut out the high-frequency ($\omega \gg \tau_c$) components of lattice motion. As these high-frequency components are necessary for most inter- and intra-molecular relaxation in the optical and infrared region, it follows that cooling a liquid or solid sample generally narrows its spectral absorption peaks considerably. Figure 10.9 provides an example of this.

An important special case of a 'relaxation' interaction is that of spontaneous emission, which was discussed briefly in section 4.9(i). The linewidth which this interaction alone would cause is known as the natural linewidth. In practice it is only an important broadening mechanism at very high frequencies, above about 10^{16} Hz.

(iv) Collision interaction

The linewidths and shapes due to collisions of the otherwise isolated excited atom or molecule have already been considered, under the assumption that the collisions are brief and highly efficient in returning the excited particle to its ground state or otherwise altering its wavefunction. This model can be tested particularly well by the microwave spectroscopy of gases, although it requires some additional theory to take account of the fairly large number of rotational states that are in fact occupied at normal temperatures. The mean collision rate between molecules can be simply and accurately controlled at low and normal pressures by variation of the pressure, and, as predicted by the theory, the linewidth is directly proportional to the pressure. Because the concentration of absorbing molecules is also proportional to the pressure, it follows that the peak heights are independent of pressure. This means that gas microwave spectroscopists cannot increase their signals by increasing pressure; they merely lose resolution. The process is called 'pressure broadening'.

The experimental invariance of peak height with pressure is a qualitative justification of the collision theory. However, it is less satisfactory quantitatively, as calculations of collision rates from linewidths can only be reconciled with calculations of collision rates using the kinetic theory of gases by assuming rather large apparent molecular diameters for pressure broadening collisions. This reflects the fact that molecules are not hard bodies, and interact appreciably at quite substantial separations. Measurements of pressure broadened linewidths can be used to estimate long-range intermolecular interaction potentials, and thus to obtain virial coefficients in accurate equations of state for gases.

(v) Exchange processes

It sometimes happens that an absorbing species has two or more well-defined sets of states, and jumps relatively rapidly between these. An example might be a proton jumping between two different basic sites, such as an OH⁻ ion and an acid anion. The proton has a different Larmor frequency at each site, and hence different Zeeman states. Because the two Larmor frequencies are close to each other, the jumps from one site to the other are not very likely to cause changes of spin orientation, i.e. relaxation. However, they do affect the line shapes of the separate absorptions by causing the lines first to broaden and then to merge. This is discussed in a classical way in chapter 7, in the context of magnetic resonance. The quantum mechanics of such systems differs from that described previously for random collisions because of the closeness in energy and limited number of the quantum states involved. Whereas in section 4.4 we were able to assume a fresh start after each collision, we now have to allow for the fact that the time-dependence of one state is permanently correlated with that of the others. The resulting theory is complex, but fortunately gives the same result as the classical theory referred to above. Chemical exchange processes mostly occur at relatively low mean frequencies, and are, therefore, most easily detected by observing the merging of lines in magnetic resonance spectroscopy. A number of such processes are described at the end of chapter 7. In electron resonance, electron exchange between free radicals provides an upper limit to the concentration of, and hence signal that can be obtained from, the radicals without serious loss of resolution of the fine structure.

6.3   Signal to noise ratio

Mention has already been made of the importance of an adequate signal to noise (S/N) ratio in high frequency spectroscopy, where the irreducible noise was identified as the statistical fluctuation in photon count rate at a given frequency band from a weak source. In principle this difficulty can be overcome by using a stronger source. Unfortunately no such solution is available in low-frequency spectroscopy, because, although high-power sources are available at microwave and lower frequencies, use of too high a source power causes saturation. The critical importance of an adequate signal to noise ratio is shown in figure 6.2 where two actual spectral absorptions are shown, one with an S/N ratio of about one and the other, of the same peak under improved conditions of measurement, with an S/N ratio of about three.

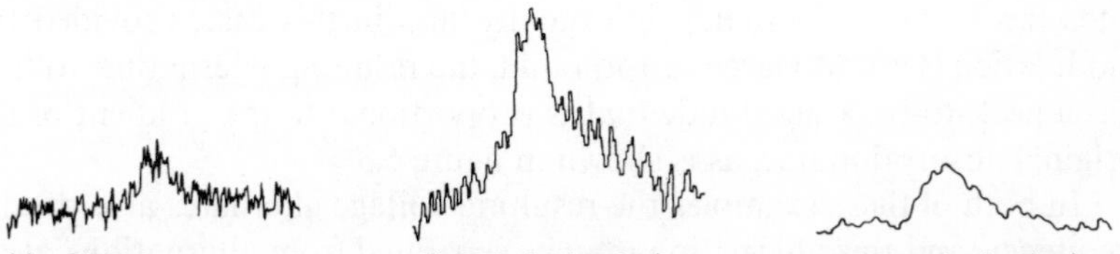

(a) S/N ≈ 1   (b) S/N ≈ 3 (signal (a)   (c) S/N ≈ 2 (signal (a)
                          enhanced by CAT)   with larger time
                                                        constant)

Fig. **6.2** Importance of adequate signal to noise ratio

A convenient definition of S/N ratio is the ratio of the maximum signal height to the r.m.s. noise height. An S/N ratio of one is necessary for any credible detection at all, and a ratio of two or three is needed for reliability and for any sort of linewidth estimation.

Noise at lower frequencies may be separated into two different types. The first is machine noise, due to spectrometer defects and source weakness. The second is Johnson noise, due to the random thermal jumps of electrons in the receiving coil or thermal detector, which produce voltage pulses that can be confused with brief signals.

(i)   Machine noise and lock-in detection

Almost any fault can cause machine noise, but many of these faults (e.g. amplifier noise) can be reduced to insignificance by good electronic design and good machine maintenance. One such design feature, the double-beam spectrometer, was mentioned in the previous chapter. A rather less tractable form of noise is that due to unsteadiness of the detector (or detectors in the case of double-beam instruments). The measured variable of a detector is typically its resistance or its impedance or its photocurrent, and all of these may be affected by mechanical shock or by unwanted changes in temperature. Fortunately a general method, called lock-in detection, is available to reduce the measured effect of such unwanted fluctuations. Lock-in detection takes advantage of the fact that if the unwanted fluctuations are analysed in terms of frequency components, they correspond to large fluctuations at or near zero frequency and very little fluctuation at other frequencies. Now a slowly swept signal is normally also detected as an almost zero-frequency fluctuation of detector output, and the combination of the two is a signal with severe baseline unsteadiness. The problem is to convert the signal temporarily into an alternating voltage at a frequency well above the maximum frequency of baseline fluctuation, and then to filter off this alternating frequency and re-rectify it. This object can be achieved in a wide variety of ways. In optical and infrared spectroscopy it is possible to use a rotating half disc, which converts the steady voltage of the signal into a square wave alternating voltage, which can be easily smoothed into a sinusoidally alternating voltage. In magnetic resonance spectroscopy the magnetic field or the frequency can be similarly modulated (i.e. made to fluctuate) sinusoidally, and in this case, provided the modulation is not too large or too rapid, the resulting alternating voltage has a peak-to-peak amplitude that is proportional to the gradient of the original absorption line, as is shown in figure 6.3.

In both of these examples the resulting voltage alternates at a precise frequency, and this alternation may be separated from alternations at all other frequencies with a sharply tuned amplifier, rather along the lines of ordinary radio tuning. This process not only cuts out the baseline fluctuations but also a good deal of Johnson noise and, as in radio, is only

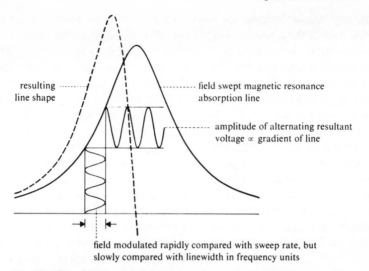

resulting ········· / ········· field swept magnetic resonance
line shape   absorption line

·········· amplitude of alternating resultant
voltage ∝ gradient of line

field modulated rapidly compared with sweep rate, but
slowly compared with linewidth in frequency units

Fig. **6.3** Effect of field modulation on a broad magnetic resonance line

limited by the need to let through those amplitude fluctuations of the
alternating voltage that occur as the signal is swept, and thus carry the
information that one is seeking. The alternating voltage is finally turned
back by simple electronics into a static voltage that drives a recorder pen.

More sophisticated spectrometers have a tuned amplifier (or other
equivalent device) that not only picks out the part of the total signal and
noise which is fluctuating at the modulation frequency, but also the part
of that frequency which is in phase (or 90° out of phase) with the original
modulation. This is known as phase-sensitive detection. The out-of-phase
component is not important at low modulation amplitudes and frequencies, but in cases where it is important it may be used to vary the
line shape from the derivative of the absorption mode to the derivative
of the dispersion mode in order to achieve better practical resolution.

The theory of modulation effects in magnetic resonance when modulation frequencies and amplitudes are high is somewhat involved,[1] for the
Zeeman states themselves are no longer constant in time. One common
application of high-frequency modulation in commercial spectrometers
is to set the modulation frequency in the kHz range in high-resolution
nuclear magnetic resonance. The alternating component is once again
selected in order to reduce spectrometer noise, and the resulting signal
appears at several new 'sideband' frequencies; with suitable phase detection the sideband component at the original + modulation frequency
looks just like the signal in the absence of modulation, bar the noise.

(ii)   Johnson noise and sensitivity limits

The root mean square voltage of the noise fluctuations produced by
random thermal jumps of electrons may be shown to be independent of

[1] J. W. Emsley, J. Feeney, and L. H. Sutcliffe (eds), *Progress in n.m.r. Spectroscopy*, Pergamon
Press, Oxford, vol. 1, ch. 1.

frequency below frequencies near to kT/h, and dependent upon the square root of the absolute temperature T at these lower frequencies. As it is not usually possible to cool the receiver appreciably, Johnson noise presents an irreducible sensitivity limitation. However, there are three ways in which its effect may be minimized. The first is to increase the signal by increasing the sample size or concentration and hence the absorption (or emission). This is normally done, although there are upper limits. In gas microwave spectroscopy the limit is set by pressure broadening or by the sheer size of the apparatus required. In electron resonance it is set by the loss of cavity Q with large samples and by Heisenberg exchange, and in nuclear magnetic resonance it is set by the dimensions of the receiver coil, which only detects resonance occurring within its own volume. The size of a tuned coil decreases as the frequency to which it is tuned increases, and this somewhat offsets the advantages of using a high magnetic field in n.m.r. Indeed it is possible to detect the resonance of protons in the earth's magnetic field by using a very large sample; this technique is of importance in geology.

The second method of reducing Johnson noise to a minimum was hinted at in the previous section. Judicious electronic 'filtering' can be used to cut out all fluctuations of the signal (considered as a static voltage) which occur above a chosen frequency. The frequency range of the noise that is permitted to pass is thus given an upper limit, which is itself chosen to fit the rate of scan of the signal. The r.m.s. noise voltage is proportional (problem 6.2) to the square root of this frequency range, and is therefore proportional to the inverse square root of the scan time, as was stated in expression 6.1. If the scan rate is too fast for the signal the recorded peak becomes flattened out and distorted (figure 6.4) with

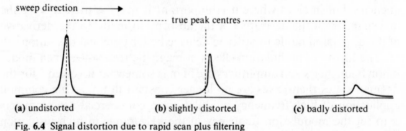

(a) undistorted   (b) slightly distorted   (c) badly distorted

Fig. **6.4** Signal distortion due to rapid scan plus filtering

resultant loss of signal and resolution. This is because the sweep through the steepest parts of the signal is equivalent to a set of superimposed sinusoidal amplitude fluctuations many of which are at frequencies which are cut out by the filter. Experimentally, the recorder pen will appear to be too sluggish to trace out the peak before it is past. In theory an infinite sensitivity may be achieved by using extremely slow sweep rates, provided that one knows in which region of the spectrum to sweep. But the limitation on this is the long term stability of the spectrometer with respect to such uncontrollable transients as mains voltage fluctuations and people who cannot refrain from leaning on it or 'adjusting' it.

The third method of noise suppression is somewhat equivalent to the second, but is more suitable in cases where a single slow sweep would easily cause saturation. It consists of sweeping the same spectrum a number, n, of times, recording it in the memory of a simple computer and then adding the n spectra to each other. The cumulative signal is of course n times as large as each individual signal, and simple random walk theory (problem 6.2b) shows that the noise is only increased $n^{1/2}$ times. Thus there is an improvement of $n^{1/2}$ in the signal to noise ratio, which is once again consistent with expression 6.1. Devices that perform this summation are called computers of average transients, or CATs.

6.4   Further sensitivity limitations

The previous section has dealt with limits to sensitivity that are most serious at very high and at low frequencies and it might be concluded that near infrared and optical spectroscopic measurements are by comparison almost infinitely sensitive. It is certainly true that they are very sensitive techniques when the resolution is good—indeed, they need to be because of the large number of absorption lines produced in this spectral region by molecules of moderate complexity. The usual practical limit on the sensitivity of these methods is the difficulty of distinguishing weak absorption or emission lines anywhere near very strong lines. Usually the chemical spectroscopist is not interested in sensitivity in order to detect single substances, but in order to detect one substance at great dilution in other substances. Alternatively he may be interested, as in Raman spectroscopy, in weak emissions very close in frequency to a very strong emission. Now a small peak is very hard to detect when it rises from what amounts to a steeply sloping baseline, particularly if the S/N ratio is poor, as may be seen in figure 6.5.

(a) flat baseline   (b) sloping baseline

Fig. **6.5** Effect of baseline slope on a peak

This practical limit becomes even more serious when the underlying strong signal is so powerful as to overload the detector or the subsequent amplifiers, or alternatively absorbs almost all the transmitted energy. In this case it is not possible to offset the sloping baseline electronically, or to use lock-in techniques in which only the desired signal is modulated. The problem is closely analogous to the way in which the eye fails to see stars during daylight hours.

A final limit to the usefulness of a spectrum is its interpretability; an absorption line is not of great interest (except for 'fingerprint' identification) unless the transition that causes it can be identified. Identification often involves taking further spectra of analogous substances, particularly isotopically substituted ones.

In recent years there has been a considerable development of new and sophisticated techniques which take advantage of the effects of applying a second radiation frequency to a sample under spectroscopic observation. These are generally described as 'double resonance' techniques. Sometimes they are used to simplify a spectrum (section 7.9 (iii)) and sometimes to induce non-Boltzmann population ratios between levels (see problem 7.3). More generally, the second applied frequency often affects only one or a few lines in a spectrum and hence helps to identify the quantum states associated with these lines.

Suggestions for further study

R. A. Sawyer, *Experimental Spectroscopy*, Dover, New York, 1963.
Laboratory spectrometer manuals. By far the best way to acquire a 'feel' for resolution, time-constants, etc. is to be able to try the effect of changing these variables on an actual spectrometer.
D. E. Clark and H. J. Mead, *Electronic, Radio, and Microwave Physics*, Heywood, London, 1961.

Problems

6.1 Two spectral absorption lines p and q have amplitudes y and peak heights h which vary with the frequency ω according to the formula

$$y = h/(1 + a^2(\omega - \omega_{p,q})^2)$$

where $\omega_{p,q}$ are the actual absorption frequencies.

a Show that the half width (at high height) of each line is $1/a$.

b Show that, when the frequency separation $\omega_p - \omega_q = 2/a$, the midpoint of the combined absorption is about 17% lower than its two maxima. Hence suggest an alternative definition of resolving power. You are recommended to substitute $x = \omega - \frac{1}{2}(\omega_p + \omega_q)$.

c Show that the derivative $dy/d\omega$ of the line in **b** shows the presence of two absorption peaks more clearly than does y.

6.2 **a** Assuming that the voltage V caused by noise in a spectrometer consists of superimposed voltage oscillations of approximately equal amplitude at discrete frequencies $n\upsilon$, where n is integral (i.e. $V \approx \sum_{n=1}^{N} a \cos 2\pi n \upsilon t$), and that the upper limit, N, is a large number, show that the root mean square noise voltage over a period of time is proportional to $N^{1/2}$.

b A drunken sailor comes out of a pub into an alleyway, and is so fuddled that he is just as likely to take a step, of length l, one way along the alley as the other way. Prove, for $N = 1, 2, 3$, and 4 steps, that his root mean square distance of travel is $N^{1/2}l$. Generalize your result approximately[2] to

---

[2] Students who like algebra may care to prove the result in the general case.

show that, if the r.m.s. noise voltage on one scan of a spectrum is V, the r.m.s. noise voltage on N accumulated scans will be $N^{1/2}V$.

c Show that doubling the resolution on a high frequency spectrometer requires a doubling of the time spent counting photons at any one frequency range in order to achieve the same signal and noise intensities. Hence show that the total spectrum scan time must be increased by four to avoid a lowering of sensitivity. Also show, using the result of **b**, that if the resolution is doubled without alteration of the scan time the signal intensity is halved and the r.m.s. noise intensity is decreased to $1/\sqrt{2}$ of its previous level.

6.3 An absorption line in a nuclear magnetic resonance experiment is found to have an unsymmetrical line shape. Suggest four possible reasons, and describe how you might attempt to find the true reason.

6.4 Evaluate expression 6.2 for hydrogen atoms at 3000 K, and hence calculate the linewidth (in Hz and nm) of the 121 nm emission line due to the Doppler effect.

6.5 Using the line shape of expression 4.18, show that the half-width at half height of a line of this shape is of the same order as the reciprocal of the average time between collisions.

6.6 Prove that the maximum height of a peak described by expression 4.18 is $\tau_0/2$, and hence that the peak heights in the microwave spectra of gases are independent of pressure. Prove also that the Lorentzian linewidth is $2/\tau_0$, and is hence proportional to the pressure.

6.7 a Show that if a spectrometer source emits a radiation field $a \cos \omega t$, and the frequency ω is modulated slightly so that $\omega = \omega_0 + b \cos \omega_m t$, where b is very small, then this is equivalent to the source providing additional frequencies $\omega \pm \omega_m$.

b Show that, if the two Lorentzian lines (centred at $\omega \pm \omega_m$ and of widths $1/a$, as in problem 6.1) are superimposed with *opposite* signs and if $1/a \gg \omega_m$, the shape of the resultant line approximates closely to the derivative of one of the line shapes, and the peak-to-peak height of the derivative shape is proportional to ω_m. Why do you think that this derivative shape is the same as the derivative shape in figure 6.3?

7 Nuclear magnetic and electron resonance

7.1 Magnetic properties of nuclei and electrons

When the Schrödinger equation was studied in chapter 2, one of the important conclusions was that all isolated systems not involving spin may only have angular momenta of $\hbar\sqrt{l(l+1)}$, where l is integral. A further conclusion was that the z-component of this angular momentum can only have values $m\hbar$, where $-l \leqslant m \leqslant l$, and m is also integral. When the additional fact of intrinsic spin was considered, the additional possibility of l being half integral was introduced. In this case m was also half integral.

This quantization of angular momentum is a very remarkable fact. The underlying unit of $\hbar$ ($= 1.054 \times 10^{-34}$ J s or 1.054×10^{-27} erg s) is the same for all bodies, whether they be electrons, nuclei, whole atoms, molecules, golf balls, or planets. Angular momentum is such a fundamental property that it is quantized in units which are quite independent of the size, mass, or moment of inertia of the body.

As a result of this, it is possible to make some accurate predictions about the angular momentum of a nucleus whilst knowing almost nothing about the shape of the nucleus or about the forces binding the nucleons together. We can for example predict that if any nucleus is subjected to Stern and Gerlach's experiment (for example in an atomic beam where the atoms have no net electronic angular momentum) it can only be deflected in a limited number of directions corresponding to the allowed values of the z-component of its nuclear angular momentum. We do, however, need further information to predict the number of allowed z-components and the magnetic moment μ associated with the nucleus.

The number of allowed z-components is always $2l+1$, where the total angular momentum is $\hbar\sqrt{l(l+1)}$. The problem is to find l for a given nucleus. This problem is closely analogous to the problem of calculating the total angular momentum of the electrons in an atom, discussed in chapter 10. Although the calculation can be done quite successfully for many atoms, it can only be done semi-empirically for nuclei because so much less is known about intranuclear forces. Certain guidelines do exist, however.[1] One is that nuclei have a shell structure similar in many respects to the electron shell structure in atoms. Just as atoms with 2, 10,

[1] A simple account is presented in C. Sharp Cook, *The Structure of Atomic Nuclei*, Van Nostrand/Reinhold, Princeton, N.J., 1964.

18, 36, 54, and 86 electrons are especially stable (inert gases), so too nuclei with 2, 8, 20, 50, 82, and 126 nucleons have unusual stability. Because the nucleons may be either protons or neutrons, the prediction of total nuclear spin[2] is made harder, but many nuclear spins have been successfully predicted. In particular, if the mass number is odd, the spin must be half integral, and if the mass number and the atomic number are both even, the spin will be zero. A nuclear spin is normally described by the quantum number I, which may be substituted for l in the more general formulae in order to predict nuclear angular momentum. Some typical nuclei with spin are listed in table 7.1.

Table 7.1

Nucleus	I	Total magnetic moment μ^*
^{1}H	1/2	24.42
^{2}H	1	6.12
^{11}B	3/2	17.52
^{13}C	1/2	6.14
^{14}N	1	2.88
^{17}O	5/2	−11.31
^{19}F	1/2	22.98
^{27}Al	5/2	21.74
^{31}P	1/2	9.89
^{51}V	7/2	29.42
^{195}Pt	1/2	5.25
^{199}Hg	1/2	4.37
(free electron)	(1/2)	(−16057.0)

* in units of $10^{-27}\,\mathrm{J\,T^{-1}}$ or $10^{-24}\,\mathrm{erg\,gauss^{-1}}$

Some common nuclei such as ^{12}C and ^{16}O have zero spin and hence no magnetic moment. As will be seen from table 7.1, there is no obvious connection between spin and total magnetic moment; the latter may have either sign relative to the spin angular momentum direction.

Nuclear magnetic moments are often described in terms of their magnetogyric ratio γ, defined by the equation $\mu = \gamma\hbar\sqrt{I(I+1)}$. Thus γ is the ratio of the total magnetic moment to the total angular momentum. Similarly electron magnetic moments are often described in terms of their g-values, where $\mu = g\mu_\beta\hbar\sqrt{J(J+1)}$ and J is the total angular momentum quantum number. For a free electron $g = 2.0023$, $J = \frac{1}{2}$, and $\mu_\beta\hbar = 9.273 \times 10^{-24}\,\mathrm{J\,T^{-1}}\,(= 9.273 \times 10^{-21}$ ergs per gauss). The spin magnetic moment of the electron, unlike that of any nucleus, may be precisely calculated using relativistic quantum theory. This theory shows that $\mu_\beta\hbar$ (known as the electronic Bohr magneton) equals $e\hbar/2m_e c$, where m_e is the electron mass and e its charge. Hence the electron μ is in the opposite direction to its angular momentum.

[2] The word 'spin' is commonly used to describe the angular momentum of nuclei, even though orbital motion may also be involved.

7.2 Nuclear and electron magnetic resonance

If a nucleus of spin I has a z-component of angular momentum $\hbar m_I$, its z-component of magnetic moment will be $m_I \mu / \sqrt{I(I+1)}$, and therefore its potential energy in an external magnetic field[3] of magnitude B_0 in the z direction will be $m_I B_0 \mu / \sqrt{I(I+1)}$. Thus $2I+1$ energy levels are available to the nucleus in the field, spread symmetrically about zero and of separation proportional to the field. Transitions may be induced between these levels by the magnetic component of radiation of the appropriate frequency. There is a selection rule $\Delta m_I = \pm 1$, which may be derived from the transition moment in exactly the same way as the $\Delta m = \pm 1$ selection rule for electrons described in chapter 10. Therefore only jumps in energy of $B_0 \mu / \sqrt{I(I+1)}$ are possible. This energy, for a proton, corresponds to a frequency of 42.58 MHz in a field of one tesla (10^4 gauss).

Exactly the same principles apply to the free electron. As its spin is $\frac{1}{2}$, it has two possible levels separated in energy by $2B_0\mu/\sqrt{3}$, which corresponds to a frequency of 27.99 GHz in a field of 1 T. As one tesla is a typical external field for a magnetic resonance experiment, it will be seen that electron resonance requires microwave radiation, with the accompanying special electronic techniques, whereas nuclear resonance requires radiation from the radio frequency range known as VHF.

So far we have only mentioned the free electron. But in a real atom or molecule there will be other interacting electrons. Furthermore any electron in a central potential may also have an orbital angular momentum, and any molecule may have rotational angular momentum. In such cases, it is often necessary to consider the quantization of the total angular momentum of the atom or molecule rather than that of the individual electron. Such considerations are of great importance in the electron resonance of transition metal ions and of isolated rotating molecules in the gas phase. Fortunately, however, many systems exist where the electron behaves magnetically as if it were virtually free. Experiments on such systems are grouped under the title of electron spin resonance and are generally easier to interpret than electron resonance experiments involving angular momentum contributions from sources other than spin.

7.3 The resonance process

(i) Classical viewpoint

So far in this chapter magnetic resonance transitions have been treated as simple energy jumps between orientational quantum states. There is, however, a useful classical view of magnetic resonance. The theory of gyroscopes shows that any magnetic moment μ will precess about a magnetic field at a frequency known as the Larmor frequency. If this

---

[3] Or, more precisely, flux density B. 'Field' is still very widely used to mean flux density, and will be used in this sense in this chapter.

precession were viewed along the x or y axes it would look like an oscillation at the Larmor frequency, with an obvious similarity to the classical picture of oscillating electrons in atoms. Such an oscillating magnetic moment would be sensitive to a magnetic field oscillating at a similar frequency at right angles to the external field. It would show absorption and dispersion in the same way as in the classical theory described in chapter 3. The Larmor frequency may thus be identified with the frequency necessary to induce a magnetic resonance transition.

The classical model above, applied to an individual nucleus, must be treated with caution. If, however, it is extended to considering the net magnetic moment of a large number of nuclei, it becomes entirely valid physically, and provides a very valuable insight into magnetic resonance experiments, particularly rather specialized ones carried out with large pulses of radiation.[4] This is because it provides a means of describing not only the net z-component of magnetic moment but also the component in the xy plane, which becomes significant when the applied radiation is near the Larmor frequency. Quantum mechanically, one might say that the spin eigenstates are no longer those appropriate to a single fixed field B_0, but are instead those appropriate to this field plus the oscillating magnetic field of the radiation. The change is substantial.

(ii)   Quantum viewpoint

We now return to a more detailed consideration of a simple two-level system, as in figure 7.1.[5] Such a two-level system is an adequate model

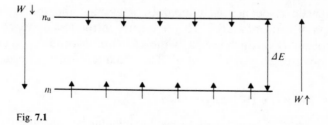

Fig. **7.1**

for all simple n.m.r. and electron resonance experiments, because even if more levels are available only $\Delta m = \pm 1$ transitions are normally allowed. Let the population of the 'upper' level (higher energy) be n_u and of the lower level n_l, with an energy separation ΔE. If thermal equilibrium of the spins is reached, then Boltzmann's law holds, and therefore

$$n_u/n_l = e^{-\Delta E/kT}$$

[4] This approach is treated in some detail in J. A. Pople, W. G. Schneider, and H. J. Bernstein, *High Resolution Nuclear Magnetic Resonance*, McGraw-Hill, New York, 1959; and in G. E. Pake, *Paramagnetic Resonance*, Benjamin, New York, 1962.
[5] The argument of this section is based on the approach in C. P. Slichter, *Principles of Magnetic Resonance*, Harper and Row, New York, 1963.

How does this thermal equilibrium come about? The spins are contained in some arrangement of atoms and molecules, liquid, solid, or gas, described by the general term 'lattice'. At any temperature above zero the particles in the lattice will occupy not only their lowest quantum levels but also, in part, higher levels, and will exchange energy by various means. The spins in the lower level are able to absorb some of this exchanging energy, provided it corresponds to the correct frequency for the Zeeman transition. But of course only the fraction of the lattice whose energy is above ΔE is capable of providing a quantum of energy ΔE. This contrasts with the situation of the spins in the upper state, which can give up ΔE to any part of the lattice. The exchange of energy thus gives the spins an 'upwards' transition probability of $W\uparrow$ and a slightly greater 'downwards' transition probability of $W\downarrow$.

At equilibrium there must be equal numbers of spins going 'up' and 'down', and therefore

$$n_l W\uparrow = n_u W\downarrow \tag{7.1}$$

so that n_l at equilibrium is slightly greater than n_u. Furthermore the average time spent in each state is $1/W$, where $W \simeq W\uparrow$ and $W\downarrow$, and so the average time that elapses before any one level gains or loses a spin is $1/2W$. This average time is known as the spin-lattice relaxation time, T_1, and because of the uncertainty principle (section 1.8) it determines the linewidth of the resonance absorption in those special cases where it is not outweighed by other broadening interactions. It may be calculated in cases where the lattice motions are known. However it is important to note that other broadening interactions normally occur (section 7.4(ii)).

Magnetic resonance is induced by subjecting the spin system to radiation of frequency $\Delta E/h$. This frequency introduces an extra transition probability w which is exactly the same for absorptive and emissive transitions (see chapters 3 and 4). The condition for equilibrium now becomes

$$n_l(W\uparrow + w) = n_u(W\downarrow + w) \tag{7.2}$$

If $w \ll W\uparrow$ and $W\downarrow$, then n_l and n_u will remain unchanged, and there will be a net upward transition rate of $(n_l - n_u)w$. The extra population of the upper level will soon fall back to the lower level because of W, and thus energy will be absorbed by the lattice at a rate $(n_l - n_u)w\,\Delta E$ per second. But $n_u/n_l = e^{-\Delta E/kT}$, and therefore

$$(n_l - n_u)w\,\Delta E = n_l(1 - e^{-\Delta E/kT})w\,\Delta E \simeq n_l(\Delta E)^2\, w/kT \tag{7.3}$$

This equation tells us that the rate of absorption of energy (and hence the area of the spectral absorption peak) in a normal magnetic resonance experiment is proportional to the number of resonating spins present ($\simeq 2n_l$), to the intensity of radiation ($\propto w$), and to the square of the energy separation of the levels and hence of the external magnetic field. It is also inversely proportional to the temperature. The rate of absorption of energy is proportional to the area of the spectral absorption peak, (although many spectrometer arrangements do not measure the energy absorption directly).

These factors are reflected in the experimental arrangements discussed in chapter 5. ΔE is very small for nuclei even in the largest stable homogeneous fields obtainable, and hence the available signal to noise ratio is often low. It is clearly of great advantage in n.m.r. to maximize the external field. The largest fields are now obtained with superconducting magnets: a few $4\,T$ ($40\,000$ gauss) high resolution spectrometers are in operation. However, high fields are less important for electron resonance as ΔE is much larger anyway, and hence signal to noise ratios are generally greater.

Up to a certain point the size of an n.m.r. or e.s.r. spectral absorption peak may be increased simply by increasing w. But there comes a point where w is no longer much less than $W\uparrow$ or $W\downarrow$, and at this point n_1 and n_u will no longer have their Boltzmann values. If $w \gg W\uparrow$ and $W\downarrow$, equation 7.2 will reduce to $n_1 w = n_u w$ (so far as the levels themselves are unaffected by the strong radiation) and hence there will be no population difference and no net absorption of radiation. This condition is known as 'saturation'. A Dutch worker failed to allow for this possibility in 1944 and thus failed to detect n.m.r. one year before the successful experiments of Purcell, Torrey, and Pound and of Bloch, Hansen, and Packard.

7.4   Time-dependent processes

(i)   Spin-lattice relaxation

It was mentioned above that T_1 was related by the uncertainty principle to the linewidth when it is particularly short. It is also possible to understand this, along with other time-dependent effects, by considering the classical model described in the previous section. In this model, the magnetic moment precesses at the Larmor frequency. Now if it has a probability $W\,(= 1/2T_1)$ of changing its Zeeman state, then its Larmor precession will be broken up into uncorrelated sections of duration $2T_1$. These may be analysed in exactly the same way as the limited wave pulse in section 1.6 showing that the precession frequency no longer has an exact value but is instead only defined within approximately $\pm 1/2T_1$ Hz. In other words the resonance is no longer sharp, but has instead a finite width of the order of $1/T_1$.

Spin-lattice relaxation comes about when the spin system exchanges energy with the lattice. Some components of the lattice exchange energy more efficiently with spins than do others. The magnetic moments of paramagnetic molecules are particularly effective in causing nuclear spin-lattice relaxation because they interact strongly with the magnetic moment of the nuclei, and for this reason the n.m.r. signals from paramagnetic substances are usually very broad. T_1 is in this case the main determiner of the linewidth.

The spin-lattice relaxation of electrons is very easily brought about by the distortions of the free electron orbital due to molecular collisions: the spin motion and the orbital motion are linked by spin–orbit coupling. Electron resonance signals are often 10^6 or more times as broad as n.m.r. signals, particularly with transition metal ions where spin–orbit coupling is large. Fortunately they are also very much stronger in intensity.

T_1 depends on temperature because the number and size of the energy jumps that take place in the lattice also depend on temperature. The energy jump distribution in the lattice can be represented by a curve somewhat analogous to the black body radiation curve (figure 1.1), and from this one may calculate the variation of T_1 with temperature.

(ii)   Spin-spin relaxation time

Other line-broadening possibilities exist which do not involve spin-lattice relaxation. The most obvious of these is inhomogeneity of the external field. If this is more inhomogeneous than the natural linewidth then the overall resonance signal will in fact consist of a number of overlapping signals from different parts of the sample, and its overall width will measure the average field inhomogeneity.

An even more drastic broadening is caused in solids by spin magnetic moments close to any one spin. A proton in ice, for example, is within 0.5 mm of about twenty other protons, all of which have magnetic fields of their own which can combine to add or subtract as much as a gauss (10^{-4} T) to or from the external field at the original proton, as described in section 7.6. The overall extent of line broadening, measured as a frequency spread Δv, is often described in terms of a second 'spin-spin' relaxation time $T_2 \equiv 1/2\pi\Delta v$. This relaxation time T_2 is just a way of describing the linewidth. Its use stems from the classical approach to n.m.r. (see footnote 4). When the linewidth is determined by spin-lattice relaxation then $T_2 = T_1$.

Fortunately both of the above broadening possibilities may be minimized by ensuring that any one spin does not stay in one particular magnetic environment for long, but instead rapidly samples all or most of the possible environments. To reduce inhomogeneity broadening it is necessary to spin the sample so that any one part of it rapidly experiences a range of external fields, both greater and less than the mean field. To avoid dipolar broadening it is necessary to set the neighbouring dipoles into rapid tumbling motion by melting or dissolving the sample. These processes are usually described as 'averaging out' the external fields. They may be understood on the classical model as follows.

The Larmor precession of a nucleus about the z axis, seen from along the x or y axes, appears as a sinusoidal oscillation of the axis of rotation whose frequency depends on the external field. If this field starts to fluctuate, the oscillation of the magnetic moment changes in frequency to fit the changing field, and at the same time becomes less well defined in frequency. When the rate of field fluctuation approximately equals the Larmor frequency, a graph of the motion of the magnetic moment with time might look as in figure 7.2.

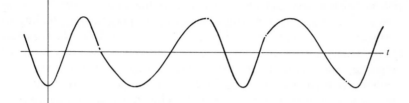

Fig. **7.2** Composite waveform containing two different frequencies

Of course the field fluctuations and hence the frequency variations are highly exaggerated in this figure. But it is easy to see that the resultant wave motion, although not sinusoidal, does approximate to a sinusoidal motion whose frequency is intermediate between the extremes depicted. It is also easy to see that if the field fluctuations were very rapid the irregularities in the wave motion would be much less obvious, and figure 7.2 would then approximate very closely to the sine wave appropriate to the average field. The remaining irregularities would lead to a slight residual uncertainty in the Larmor frequency which would give the resonance line a finite width measurable, as before, in terms of a time T_2.

The conclusion of all this is that really narrow n.m.r. lines are only obtained with liquid or dissolved samples contained in a tube spun about its axis. This is the normal experimental arrangement in n.m.r., and the lines so obtained are commonly 10^3 or 10^4 times narrower than in solids. Such precautions are not, however, usually necessary in electron resonance as the T_1 broadening is so much greater anyway. The measurement of T_2 (as a linewidth) at different temperatures may reveal processes such as the onset of free or partially free rotation of certain ions, such as NH_4^+, in a solid.

(iii) Exchange processes

There are other ways in which time-dependent processes are of great importance in magnetic resonance. Spins often have slightly different Larmor frequencies in different environments (see section 7.6). But if they exchange rapidly between these environments their n.m.r. spectrum will be a single line corresponding to their average Larmor frequency. For example, the protons in aqueous hydrochloric acid jump rapidly between Cl^- ions and H_2O molecules. They therefore give rise to a single resonance line at a field which is the average of the fields necessary to produce resonance in pure HCl, pure H_2O, and pure H_3O^+. The average is, of course, weighted by the relative amounts of each species present in normal equilibrium.

If the rate of proton exchange could be drastically slowed, the resonance line would begin to broaden as the Larmor precession component became less and less sinusoidal. If the rate of exchange were slowed even more, three separate lines would appear corresponding to the independent proton resonances of HCl, H_2O, and H_3O^+. The amount of broadening in a given system may be accurately related to the rate of exchange. Approximately, maximum broadening is observed when the mean rate of exchange equals the inverse of the slow exchange limit peak separation measured in frequency units. Thus magnetic resonance may be used in favourable cases for kinetic studies of exchanges occurring more rapidly than once per second—a region not easy to observe by other methods. An example is given in problem 7.13, and in figure 7.3.

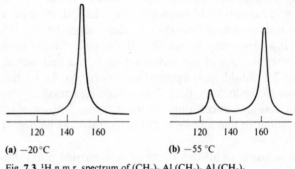

(a) −20 °C   **(b)** −55 °C

Fig. 7.3 ^1H n.m.r. spectrum of $(CH_3)_2$ Al $(CH_3)_2$ Al $(CH_3)_2$ showing bridging—terminal CH_3 exchange at −20 °C [6]

Other uses include studies of axial-equatorial chair transitions in substituted cyclohexanes, studies of restricted rotation, and studies of rapid ligand exchange with paramagnetic transition metal ions. The ability of n.m.r. to detect molecular conformations lasting for only a fraction of a second in the liquid state has opened up the new field of conformational analysis, and has led to greater understanding of organic reaction mechanisms.

7.5   Simple resonance experiments—single line observed

Measurement of γ and g factors

A single magnetic resonance experiment leads to a value of $g\sqrt{J(J+1)}$ or of $\gamma\sqrt{I(I+1)}$. Values of γ are of interest to theoretical physicists developing theories of nuclear structure. Values of g and J are useful in inter-

[6] E. A. Jeffrey, T. Mole, and J. K. Saunders, *Chem. Comm.*, 696, 1967.

preting the electronic structure of many free radicals, because the g-value of most free radicals differs slightly from the free electron value, owing to the interaction of the unpaired electron with other electrons. g-values may sometimes be used to distinguish different free radicals. A recent application of this has been in cancer screening research; it has been shown that cancerous and non-cancerous tissue both give broad single-line electron resonance spectra, but with significantly different g-values.

When single crystals of a substance are studied the g-value can in fact depend on the orientation of the trapped free radical relative to the magnetic field. In this case a knowledge of the variation of g with radical orientation gives information about the symmetry of the radical. For example, radical ions CO_3^- may be created by irradiating a single crystal of $KHCO_3$. The CO_3^- radicals are trapped in the positions of the parent HCO_3^- ions, and thus their orientations in the crystal are known. If the host crystal is rotated so that the magnetic field is always parallel to the CO_3 plane, the observed g-value of the electron resonance is not constant (constancy would be expected for a symmetrical ion with all OCO bond angles equal to 120°) but instead varies slightly, showing two maxima and two minima in a 360° rotation. Hence it is apparent that the CO_3^- does not have the full symmetry of the CO_3^{2-} ion but is distorted (see figure 7.4).[7]

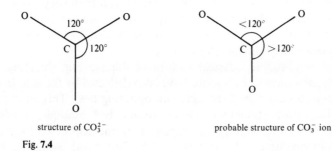

structure of CO_3^{2-} probable structure of CO_3^- ion

Fig. **7.4**

Measurement of B_0

If the magnetogyric ratio of a nucleus is known, the frequency at which resonance occurs in a given field measures the field. This principle is used in several magnetometer devices, including ones which are suspended beneath surveying aeroplanes to measure the earth's magnetic field. Unusual variations in the earth's field in particular places are of interest to geo-prospectors.

Measurement of concentration

The area under a resonance signal is proportional to the number of spins undergoing resonance. This fact may be used to measure concentrations,

[7] G. W. Chantry, A. Horsfield, J. R. Morton, and D. H. Whiffen, *Molecular Physics*, Vol. V, page 589, 1962.

by comparing the resonance signal areas of known and unknown substances. There is often no other reliable way of measuring free radical concentrations. NMR is sometimes used to measure the free water or hydrocarbon content of pastes and slurries.

7.6 Measurement of chemical shift

The great usefulness of n.m.r. in organic chemistry relies above all else on the fact that in an atom or compound the magnetic field B necessary to cause resonance at a given frequency varies slightly from the field B_0 appropriate for a free nucleus. The relative difference, $\sigma = (B_0 - B)/B_0$ is known as the shielding constant, and measures the diamagnetic screening effect of the electrons close to the nucleus. The shift in the resonance conditions is known as the chemical shift,[8] and is normally measured in terms of δ, where

$$\delta = 10^6(\sigma_{TMS} - \sigma_{sample}) = 10^6 \frac{(v_{sample} - v_{TMS})}{v_{TMS}} \text{ at constant } B_0$$

TMS stands for tetra-methyl silane, a convenient reference compound with a low shielding constant; for obvious reasons the free proton is not a convenient reference, although its σ is precisely zero.

δ is normally measured directly in commercial frequency-swept spectrometers, because these have chart paper whose rate of motion is linked to the frequency sweep mechanism, and which is calibrated precisely for the particular B_0 of the spectrometer.

There is a simple classical picture of diamagnetic shielding. The electrons surrounding the nucleus have been deflected by the field in such a way as to produce (by Lenz's law) an opposing field. This amounts to a shielding, proportional in amount to B_0. For example, a pair of electrons in a p-orbital might interact so that one acquired a little extra angular momentum and one lost a little. The result would be a net magnetic moment opposing the external field.

Protons rarely show shifts of more than 10 ppm (i.e. 10δ). However, shifts may readily be measured to 0.01 ppm. Also, protons in any one type of functional group usually have rather similar chemical shifts. Therefore the observed chemical shift of a proton is a very useful clue as to its chemical environment. The likely proton shift range for certain common functional groups is illustrated in figure 7.5.[9] There are a large number of semi-empirical theories to explain these shifts. For example, a decrease in electron density around a proton, due to its being attached to an electronegative group, generally leads to a reduction in shielding and hence to a 'downfield' or 'high frequency' shift. A fully reliable calculation of a chemical shift, however, requires an accurate knowledge

[8] Chemical shifts are sometimes quoted in terms of their τ values, where $\tau = 10.00 - \delta$.
[9] A very extensive list of chemical shifts is published by Varian Associates, Palo Alto, California (NMR Spectra Catalog, pub. 1962).

R=alkyl group or H

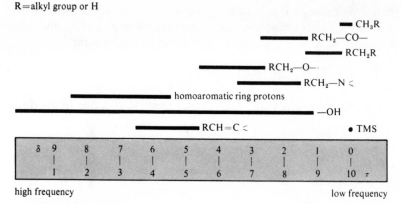

Fig. **7.5** Typical proton chemical shifts

of all the electronic wavefunctions and is therefore not possible except for hydrogen. One difficulty in comparing experiment with theory is that in most molecules the diamagnetic screening of the nucleus depends on the orientation of the molecules with respect to the magnetic field. The single chemical shift value measured with a liquid sample is in fact the average chemical shift over all orientations, the different resonance frequencies being averaged by rapid tumbling.

The chemical shifts observed in resonance experiments on heavier nuclei are usually much larger than those observed in proton resonance experiments, and may be greater than 1000 ppm. This is partly because of the far greater number of electrons involved in diamagnetic screening, and partly because other chemical shift mechanisms become important as atomic number increases. The best known of these is temperature independent paramagnetism,[10] which causes shifts to high frequency and arises from the physical distortion of the electron orbitals in the external magnetic field.

An example of the direct usefulness of the chemical shift phenomenon was afforded by a study of the ^{205}Tl resonance in a melt of empirical formula $TlCl_2$.[11] Two ^{205}Tl resonances of equal area were observed, separated by 2480 ppm, and it was concluded that the correct formula was probably $Tl^+ TlCl_4^-$.

7.7   Dipolar coupling in solids

The broadening of resonance lines in solids due to the various additional local fields produced by neighbouring spins was discussed in section 7.4. We now consider this dipolar interaction in more detail. Spin dipoles in a magnetic field point on average either along or against the field, as their components in other directions average out due to Larmor pre-

[10] Not to be confused with ordinary paramagnetism, which may produce such large shifts and broadenings as to render the resonance unobservable.

[11] T. J. Rowland and J. P. Bromberg, *Journal of Chemical Physics*, **29**, 626, 1958.

cession. The energy of interaction E between two parallel dipoles centred at points A and B whose moments have magnitude μ_a and μ_b, and which are separated as shown in figure 7.6, is given by the formula[12]

$$E = 10^{-7} \frac{\mu_a \mu_b}{r^3} (3 \cos^2 \theta - 1) \tag{7.4}$$

E is in joules and μ is in JT^{-1}, and hence the field at A due to the dipole

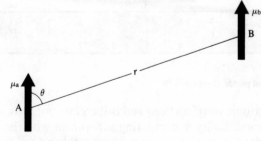

Fig. 7.6 Dipolar interaction

at B is $10^{-7} \mu_b (3 \cos^2 \theta - 1)/r^3$ tesla. Equation 7.4 may be made somewhat less abstract by considering the field produced at a proton by another proton at a fixed distance of 0.16 mm (as in a CH_2 group) as the proton pair is rotated relative to the field. The variation of this 'internal' field is shown in figure 7.7. It reaches a maximum of 0.70×10^{-3} T (7 gauss) and a minimum of 0.35×10^{-3} T in the opposite direction. It might appear from this figure that a positive internal field is more likely than a negative

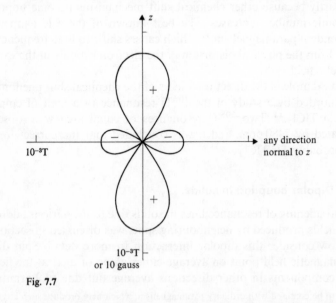

Fig. 7.7

---

[12] The 10^{-7} in this formula arises in the use of SI units from Ξ, the magnetic permeability constant, which is necessary in order to make equation 7.3 dimensionally consistent.

one for parallel protons. However, the figure should really be in three dimensions. It would then be easier to see that the positive and negative regions are of identical surface area. Mathematically, $\int(3\cos^2\theta-1)\,d\tau = 0$, and hence the dipolar field at A averages to zero when the point B samples all angles θ due to rapid molecular tumbling, as in liquids. It also averages to zero when the dipole at B arises from an electron spin in an s-orbital, which has no preferred direction in space.

There are two important cases where the dipolar field averaging described above does not take place. The first, of importance in n.m.r., is with solids. The resonance line in solids may be thought of as a super-imposition of the resonance lines of the nuclei under study in the various total fields produced by adding the internal dipolar fields to the external field. In some cases the nearest spins (which contribute the greatest part of the internal field because of the r^{-3} term in equation 7.4) have a simple arrangement around the nucleus under study, such as making the second and third vertices of an equilateral triangle. An example might be the protons in a methyl group. The resonance of the solid then has an apparent structure instead of the ordinary broadening expected for a random arrangement of spins. Figure 7.8 shows examples of this structure

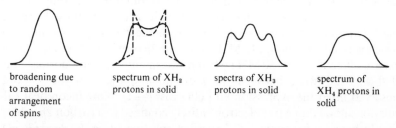

broadening due
to random
arrangement
of spins

spectrum of XH_2
protons in solid

spectra of XH_3
protons in solid

spectrum of
XH_4 protons in
solid

Fig. **7.8** Solid state proton n.m.r. spectra

for proton resonances of the groups XH_2, XH_3, and XH_4. The dotted lines show the field spread expected for the isolated XH_2 unit. The real resonance is further broadened and hence smoothed by the fields of more distant spins. An example of the application of this was in a study of 'infusible white precipitate', produced by adding ammonium hydroxide to aqueous mercuric chloride. The proton resonance spectrum of the solid showed that the protons were grouped in pairs (XH_2 above) and therefore that the original ammonium ions had reacted to form $-NH_2$ groups and were not present as NH_3 or as NH_4^+.

A second example of incomplete averaging occurs in electron reso-nance when an unpaired electron in an orbital other than an s-orbital interacts with the spin of the nucleus at the centre of the orbital. If the free radical is trapped in a solid, the electron's orbital will be fixed in direction relative to the external field, and therefore the electron will not be free to travel unrestrictedly in all the regions of space surrounding the nucleus. Its dipolar field at the nucleus will therefore not average to zero, but instead average to a value dependent on the orientation of the radical to the field. Conversely, a related orientation dependent field shift will occur in the electron resonance due to the field of the nucleus. The electron resonance will in fact be split into as many equally spaced resonances as there are allowed nuclear spin z-components. This splitting will add to or subtract from the hyperfine splitting discussed in the next section, but may be readily distinguished from it by its dependence on the orientation with respect to B_0 of the crystal containing the free radical. For a given atom the size of the various possible atomic orbitals is approximately known, as are the electron and nuclear magnetic moments. Thus, with the aid of equation 7.4, the electron resonance splitting may be calculated for the situation where the orbital is occupied by one unpaired electron. By comparison of this calculated splitting with the observed splitting it is possible to estimate the percentage of time spent by the free electron in the orbital under consideration (or, more precisely, to estimate the contribution of that orbital to the overall wavefunction of the unpaired electron).

This may be illustrated by describing a study of the electron resonance spectrum of the radical $^{13}CH(COOH)_2$, obtained by irradiation of isotopically substituted malonic acid and trapped in a specific orientation in the host crystal. The unpaired electron was found to be mainly located in a carbon $2p$ orbital normal to the plane in which the three bonds to the ^{13}C atom lay. This was deduced from the ^{13}C hyperfine splitting, which varied from about 210 MHz when the $2p$ orbital was parallel to the external field to about 30 MHz when it was in any direction normal to it. If θ was the angle between the p-orbital and the field the splitting observed fitted the expression $90 + 60(3\cos^2\theta - 1)$. Now theoretical calculation shows that a free electron entirely confined to a carbon $2p$ orbital should have an isotropic ^{13}C hyperfine splitting of ca. $90(3\cos^2\theta - 1)$. The same electron entirely confined to a carbon $2s$ orbital should have an isotropic splitting of ca. 3000 MHz (see next section). Thus the observed anisotropic hyperfine splitting suggested a negligible occupation by the unpaired electron of the carbon $2s$ orbital, but a $\frac{2}{3}$ occupation of the $2p$ orbital. The remaining $\frac{1}{3}$ occupation was deduced to arise from delocalization of the unpaired electron onto the carboxyl groups.

Much approximate information of this kind has been deduced from electron resonance studies of species produced by irradiation of a host crystal. The result has been to extend the known range of small molecules and ions to include species such as SO_4^-, PO_4^{2-}, F_2^-, and XeF.

7.8 Scalar, spin-spin, and hyperfine coupling

(i) Hyperfine coupling

Equation 7.4 in the previous section is derived by considering two dipoles whose length is very short in comparison to their separation. This is a valid approximation in all cases except that of an electron in an s-orbital around a nucleus. However, the wavefunctions $\psi(r)$ of all s-orbitals are infinite at $r = 0$, and therefore the electron density $r^2\psi^2(r)$ is not vanishingly small within the region of the nucleus. Therefore in addition to the long-range dipolar interaction described by equation 7.4 there is an important short-range interaction, known as Fermi contact. Analysis shows that it is proportional to the interaction $\mu_n\mu_e \cos \theta$ that would be expected for two classical bar magnets with moments μ_n and μ_e overlapping directly at an angle θ as in figure 7.9. The most important

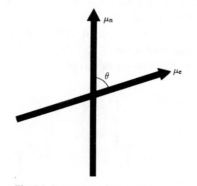

Fig. **7.9** Fermi contact interaction

experimental consequence of Fermi contact is that the nucleus experiences an additional field in the z-direction proportional to the z-component of the electron spin magnetic moment, and conversely that the electron experiences an additional field in the z-direction proportional to the z-component of the nuclear magnetic moment. The electron resonance therefore occurs at $2I+1$ uniformly spaced values of the external field corresponding to the $2I+1$ possible z-components of the nuclear moment. This splitting of the electron resonance is another type of hyperfine structure, and the interaction is known as scalar, or isotropic hyperfine coupling. It is independent of the orientation of the radical in which it occurs, because the s-orbitals responsible for it are independent of direction, and its magnitude is determined by the s-orbital wavefunction of the unpaired electron and by the physical size and magnetic moment of the nucleus that overlaps this s-orbital.

The existence and amount of the hyperfine splitting in the electron resonance spectrum of a free radical may be used either to show the presence of a nucleus of a particular spin in the free radical, or conversely to show that the unpaired electron is delocalized into s-orbitals centred on that nucleus. One of the first experiments to detect hyperfine splitting was done by Griffiths and Owen on a dilute solid solution of Na_2IrCl_6 in Na_2PtCl_6.[13] The unpaired electron centred on the iridium ion showed hyperfine structure in its resonance spectrum due to 5% delocalization onto the chloride ligands, which have nuclear spin.

[13] J. H. E. Griffiths and J. Owen, *Proc. Roy. Soc. A*, **226**, 96, 1954.

When several nuclei interact with the unpaired electron they each contribute their internal field and hence the hyperfine splitting pattern becomes more complex. For example, suppose that a single proton in an organic free radical contributes an internal field at the electron of $\pm 5 \times 10^{-4}$ T (5 gauss). A second equivalent proton provides an additional $\pm 5 \times 10^{-4}$ T, and so the two together provide possible internal fields of $\pm 10 \times 10^{-4}$ T or of zero, the latter being twice as probable as either of the former. The electron resonance signal is thus split into a symmetrical triplet, whose peak heights are in the ratio $1:2:1$.

The spectrum of the nitrobenzene negative ion $C_6H_5NO_2^-$ is shown in figure 7.10. This ion was prepared inside the resonance cavity by reacting a flow of liquid ammonia containing nitrobenzene with a flow of a dilute solution of sodium metal in liquid ammonia. The spectrum is of

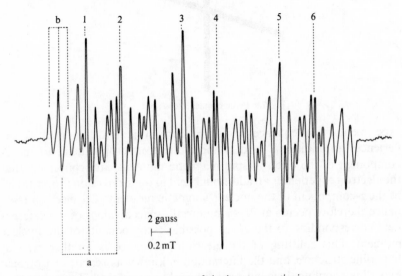

Fig. **7.10** Electron spin resonance spectrum of nitrobenzene negative ion

average complexity. The basic structure of three doublets is clear (lines 1 to 6). These arise from the $m_I = +1$, 0, and -1 states of the ^{14}N nucleus, further split by the $m = \pm\frac{1}{2}$ states of the para proton. The meta and ortho proton pairs are responsible for the remaining structure; the ortho pair splits each of lines 1 to 6 into a $1:2:1$ triplet (e.g. a) and the meta pair splits each of the resulting lines into a less extended triplet (e.g. b).

The various mechanisms by which the unpaired electron is delocalized into the s-orbital of nearby nuclei are a subject of current research. A fairly well established empirical formula exists relating the fractional π–electron density, ρ, at an aromatic carbon atom to the observed splitting a_H of the proton attached to that atom. It is $a_H = -63\rho$ MHz.

Electron resonance hyperfine splitting has provided strong evidence for the validity of Hückel π–electron theory and for the existence of hyperconjugation.[14]

The hyperfine splitting between the proton and the electron of a free hydrogen atom is the easiest hyperfine splitting to predict theoretically, and is of fundamental importance. It has a value of 1420 MHz, and may be measured either from electron resonance or from optical spectra. It may also be detected as an emission from the hydrogen atoms in outer space. In this latter case the vast size of the 'sample' compensates for the extremely low natural probability of the antiparallel to parallel spin transition.

The above description of hyperfine splitting has been in terms of the internal fields produced by the electron at the proton and vice versa. It is, however, important to emphasize that, if it is described as an energy of interaction, the splitting of any spin a by another spin b is the same as the splitting of b by a. This is why hyperfine splittings are normally quoted in energy (frequency) units rather than in field units. The reason for this lies in the symmetry of the basic interaction $\mu_n\mu_e \cos\theta$. If either spin reverses its Zeeman orientation the interaction energy simply changes sign and hence adds or subtracts an energy $2\mu_n\mu_e \cos\theta$ to the transition energy of the other spin. The same argument applies to the spin-spin splittings described in the next section.

(ii)   Spin-spin coupling

Nuclei interact with each other in a manner closely analogous to the hyperfine nucleus-electron interaction described above. Their interaction is called spin-spin coupling, and results in splittings in n.m.r. spectra which have about one millionth of the energy of hyperfine splittings, but which are nevertheless easily detected by a high resolution instrument. Spin-spin coupling arises from the same nucleus-electron interactions that give rise to hyperfine coupling in electron resonance spectra. It is, however, harder to describe or predict accurately because it involves all the electrons in the molecule rather than just the unpaired electron.

Let us consider a simplified model of a $^1H^{19}F$ molecule in a magnetic field. If the ^1H nucleus is oriented 'with' the field, it will add to the field experienced by the electron in the H—F bond which is oriented 'with' the field, and hence decrease its potential energy. Conversely it will add to the field which the other electron in the bond is 'against', and thus increase that electron's potential energy. The parallel electron is able to decrease its potential energy still further by moving on average slightly towards the proton; also the antiparallel electron is able to reduce its gain of potential energy by moving away from the proton. As a result of this the antiparallel electron will be nearer the ^{19}F nucleus, on average, than will the parallel electron, so the ^{19}F nucleus will experience non-cancelling

[14] See L. Salem, *The Molecular Orbital Theory of Conjugated Systems*, Benjamin, New York, 1966.

hyperfine interactions from the two electrons in the H—F bond which will leave it with a residual internal field. Hence the ^{1}H and ^{19}F nuclei are each slightly affected in energy by the direction of the other's spin. This constitutes spin-spin coupling.

A more complete description would take into account the other electrons in the molecule. Actual splittings may have either sign and may also occur between nuclei separated by several bonds.[15] Spin-spin splitting constants are normally measured as a frequency, equal to the observed line separation when the applied frequency is scanned at fixed field. They are, of course, independent of external field and of the angle between the bond and the external field. Typical proton-proton splitting constants are between one and fifteen hertz. However, splitting constants involving heavier nuclei are much larger because the Fermi contact is much greater. For example, splittings of over 5 k Hz have been observed between ^{31}P and ^{195}Pt nuclei in platinous organophosphine complexes.

When several equivalent nuclei all interact with another nucleus the resonance of that nucleus is split into multiplets exactly as in electron resonance. The appearance of these multiplets in an n.m.r. spectrum provides evidence for the presence of the appropriate atomic groupings. In an organic compound, for example, a quartet of lines whose heights are in the ratio $1:3:3:1$ is very likely to arise from a —CH$_3$ group near the nucleus undergoing resonance, and a $1:2:1$ triplet probably indicates the presence of a —CH$_2$ group. A typical simple n.m.r. spectrum, measured at 60 MHz, is shown in figure 7.11 to illustrate these points.

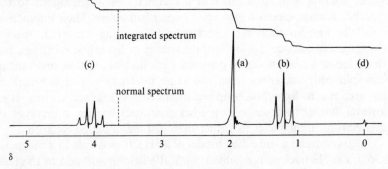

Fig. **7.11** 60 MHz spectrum of ethylacetate + TMS reference

The main features of this spectrum are (a) the singlet at $\delta = 1.95$, arising from the acetate methyl protons, whose spin-spin coupling with other protons is negligible, (b) the triplet centred on $\delta = 1.19$ and of splitting 7.2 Hz, arising from the alkyl chain methyl group split by the methylene group, (c) the corresponding quartet at $\delta = 4.06$, with the same splitting, due to the methylene group split by the neighbouring methyl protons, and (d) the small signal at $\delta = 0.00$, due to about 2% of added TMS.

[15] An extended discussion of the many theories explaining the magnitudes of spin-spin couplings is found in J. W. Emsley, J. Feeney, and L. H. Sutcliffe, *High Resolution Nuclear Magnetic Resonance Spectroscopy*, Pergamon Press, Oxford, 1965.

There are several additional features of interest. The separate line above the spectrum is made by a second scan of the resonance region using alternative electronic circuitry that plots not the peak height but the peak area traversed. The height of any one step (corresponding to the peak at the same field in the original spectrum) may be compared with that of any other step to obtain relative peak areas and hence relative numbers of protons undergoing resonance. The 3:2 area ratio of the —CH_3 and the —CH_2 groups is obvious from this integrated spectrum. The two tiny peaks at $\delta = 3.02$ and 0.86 arise from the expected 1% of $C_2H_5O.CO.^{13}CH_3$ molecules in solution, where the ^{13}C spin further splits the acetate methyl resonance. The small 'ringing' marks after each peak are a result of scanning the resonances slightly too rapidly to permit

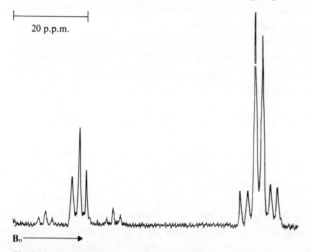

├────────────┤

20 p.p.m.

B_o ──────→

Fig. **7.12** The ^{19}F n.m.r. spectrum of $XeOF_3^+$ showing splittings due to 29% ^{129}Xe and suggesting a trigonal bipyramidal structure with two axial F's. (R. J. Gillespie, B. Landa, and G. J. Schrobilgen, *Chem. Comm.* 1972, 607.)

the full equilibrium of spins and radiation which has so far been assumed.

A second example of nuclear spin-spin splitting is given in figure 7.12. The ^{19}F spectrum shows the presence of two equivalent and one unique proton in $XeOF_3^+$, suggesting a structure

The presence of 29% $^{129}XeOF_3$ ($I = \frac{1}{2}$) is apparent.

7.9   More complex features of magnetic resonance spectra

(i)   Equivalent nuclei

Although spin-spin couplings occur between equivalent nuclei, just as between non-equivalent ones, they do not give rise to similar line splittings in simple spectra.[16] The ^{1}H resonance of benzene, for example, is a single line. This helps to simplify the interpretation of spectra. However, the physical reasons why the interactions of equivalent nuclei do not give rise to splittings are not particularly simple. Consider a hydrogen molecule. If the two nuclei are considered separately, and their spin wavefunctions are represented as α or β depending on the sign of the z-components of nuclear angular momentum, then the four overall nuclear spin states would appear to be $\alpha(1).\alpha(2)$, $\beta(1).\beta(2)$, $\alpha(1).\beta(2)$, and $\beta(1).\alpha(2)$. However, the last two states as written are not acceptable. For example, the state $\alpha(1).\beta(2)$ implies that if the two protons were exchanged the spin state would change from $\alpha(1).\beta(2)$ to $\alpha(2).\beta(1)$ or vice versa. This would be a physically detectable change which would violate the basic stipulation that the protons are indistinguishable.

Fortunately this difficulty can be avoided. The overall states $\alpha(1).\beta(2)$ and $\beta(1).\alpha(2)$ are equal in energy, and hence linear combinations of them are also solutions of the wave equation for the entire molecule. The combination wavefunctions

$$\frac{1}{\sqrt{2}}[\alpha(1).\beta(2)+\beta(1).\alpha(2)] \quad \text{and} \quad \frac{1}{\sqrt{2}}[\alpha(1).\beta(2)-\beta(1).\alpha(2)]$$

are therefore permissible. If the nuclei are now exchanged the first of these wavefunctions will be unaffected (symmetric spin state) and the second will simply change sign (antisymmetric spin state). In either case the square of the wavefunction and hence the properties of the molecule are unaffected by the exchange of nuclei. The combination wavefunctions are therefore consistent with the equivalence of the two nuclei. Thus if we stipulate that the two nuclei are equivalent we are forced to conclude that their spins may not be considered separately, but only in combination. If we were now to go on to study the effects of spin-spin coupling on the four states we would find (after a detailed calculation) that the three states $\alpha(1).\alpha(2)$, $\beta(1).\beta(2)$ and $1/\sqrt{2}[\alpha(1).\beta(2)+\beta(1).\alpha(2)]$ were all equally affected in energy, so that transitions between them occured precisely at the Larmor frequency. Furthermore, transitions from these symmetric states to the antisymmetric state cannot normally occur, for the reasons given in chapter 4, so no splittings would be expected.

The above description is accurate but abstract. It can be complemented by a pictorial but rather loose semi-classical description. A spin wavefunction such as ↑↓ is inadequate because it only describes the spin z-components. However, equivalent nuclei have the same Larmor precession frequency and hence one may not ignore or 'average out' the interactions arising from the other components of their spin magnetic moments. These other components will lead the spin moments to precess at certain fixed angles to each other as well as to the external field. Let

[16] They can, however, be detected in second-order spectra, of the type discussed in the next sub-section.

us consider the hydrogen molecule again. Because of the spin interaction and the symmetry of the molecule the antiparallel spins may be considered to precess either exactly in phase, or exactly $180°$ out of phase, as illustrated in figure 7.13. The moments of the two out-of-phase spins

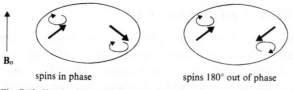

spins in phase   spins $180°$ out of phase

Fig. **7.13** Simple picture of the precession of antiparallel spins in H_2

cancel completely so that at a distance they appear to behave like a single nucleus of zero spin. The two in-phase spins have z-components of $\pm\frac{1}{2}\hbar$ and components normal to this of $\pm\sqrt{2}\hbar/2$ (i.e. a total magnitude of $\sqrt{3}\hbar/2$) and hence their combination has a z-component of zero and a total magnitude of $\sqrt{2}\hbar$—exactly as expected for the $m = 0$ state of a spin angular momentum where $I = 1$. Similarly the remaining two states, $\alpha(1).\alpha(2)$ and $\beta(1).\beta(2)$ have total z-components of $\pm\hbar$, like the $m = \pm1$ states of a spin $I = 1$, and thus three overall angular momentum states exist, closely analogous to those of a single nucleus of spin $I = 1$. Transitions between the $I = 1$ and $I = 0$ states are very infrequent, because they are spin forbidden (chapter 4), and therefore a sample of molecular hydrogen behaves as if it were a mixture of two species, one with a 'combined nucleus' spin of zero and the other, three times as common, with a 'combined nucleus' spin of one, having, of course, a single magnetic resonance absorption frequency. The former pseudo-species is called para-hydrogen and the latter ortho-hydrogen. The probabilities of occurrence and the symmetries of nuclear spin states such as these are of great importance in interpreting the rotational spectra of symmetrical molecules.

(ii) Almost equivalent nuclei

When nuclei have chemical shifts comparable in frequency units to their spin-spin interactions they are almost equivalent, and as a result their n.m.r. spectra become very complex and often impossible to interpret. This is because the nuclear spin states, as with equivalent nuclei, are not independent, while at the same time there are no strict selection rules to provide simplifications. Complex 'second order' splittings of this type are common among aromatic protons. They may often be reduced in importance by an increase in external field and hence in the chemical shift (measured as a frequency) between the nuclei.[17] However they are very

[17] A recent and fairly straightforward account of second-order splittings is found in R. M. Lynden-Bell and R. K. Harris, *Nuclear Magnetic Resonance Spectroscopy*, Nelson, London, 1969.

common and mean that the high-resolution n.m.r. spectroscopist has to be very careful in his interpretation of complex spectra. Problem 7.13 provides an example of second-order effects on peak heights, and these are also apparent in the slightly uneven doublet of figure 7.12.

(iii)   Collapse of spin multiplets

If a nucleus, such as the hydroxylic proton in ethanol, exchanges rapidly with other nuclei from different molecules it not only experiences the average chemical shift of its environment but also fails to produce hyperfine splittings at other nuclei or at unpaired electrons in the molecule. This is because the internal field experienced by other nuclei or electrons due to hyperfine interactions with the exchanging nuclei fluctuates rapidly in sign as different nuclei with different spin components successively attach themselves to the molecule, and therefore averages to zero. The spin-spin splittings due to the hydroxylic proton in ethanol, for example, are only detectable when the rate of proton exchange is drastically slowed, for example by drying and neutralization of the ethanol. This effect is illustrated for methanol in figure 7.3.

If the nuclei do not exchange, but instead relax very rapidly, an identical effect occurs. For example, ^{35}Cl and ^{37}Cl nuclei relax rapidly and hence do not produce detectable splittings in the resonances of nearby nuclei. The spin state of any nucleus may also be greatly altered (see section 7.3(i)) by intense irradiation of its own resonance, leading to collapse of spin multiplets due to that nucleus.[18] This 'double irradiation' experiment may be carried out in order to simplify complex spectra.

If free radicals approach closely their unpaired electrons can exchange with each other. Each electron then experiences the average internal magnetic field of the equivalent nuclei on each free radical it visits. Averaged over a large number of radicals this field is zero and therefore the electron resonance spectrum collapses to a single line. The electron exchange process (Heisenberg exchange) puts an upper limit on the concentration of free radicals which may be detected without significant loss of hyperfine structure information.

(iv)   Triplet states in electron resonance

Almost all the discussion of electron resonance so far has concerned measurements on isolated unpaired electrons. However, many atomic and molecular states exist in which two or more electrons are unpaired. The spin and space wavefunctions for atomic states with two parallel electrons are discussed in chapter 10; in the absence of any interaction with orbital angular momentum they are known as triplet states, because under these conditions the electrons behave like a single particle of spin $S = 1$, with three magnetic sub-levels $m_S = 1, 0,$ and -1.

[18] Less intense irradiation leads to a variety of other interesting and useful effects.

These states may be understood semi-classically in exactly the same way as the triplet combined nuclear states in section 7.9(i). In particular, the $m_S = 0$ state corresponds to a different mutual electron-electron orientation from that in the $m_S = \pm 1$ states. This is important in molecules, because if the electrons move in non-spherically symmetric orbitals then their dipole-dipole interaction (expression 7.4) will not average to zero, but will instead vary, like the electron-nucleus dipolar interaction energy described previously, as the molecular axes are rotated with respect to the external field. This interaction energy will, in an axially symmetric molecule such as coronene, shift the $m_S = 0$ level in energy, and will also shift the $m_S = \pm 1$ levels by a different energy, as shown in figure 7.14, and so the resulting spectrum will show two peaks instead of

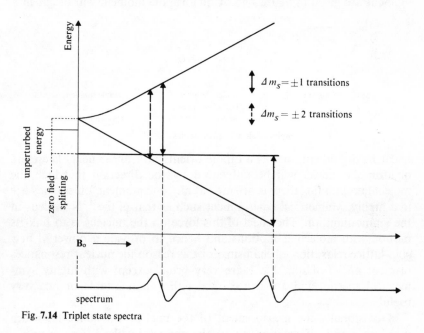

Fig. **7.14** Triplet state spectra

the one peak expected for atoms. The dipolar electron-electron interaction will also be present at zero external field, although in this case the electrons will orient in each other's field rather than in the external field. The electron states under these conditions are not strictly eigenstates of angular momentum, although they may be approximately described in terms of the quantum numbers appropriate to the high field eigenstates, and for this reason $m_S = \pm 2$ transitions are weakly permitted in molecular triplet states (see figure). The energy difference of the states at $\boldsymbol{B}_0 = 0$ is known as the zero-field splitting.

In molecules without axial symmetry, such as naphthalene, three zero-field levels are observed and the spectrum is even more complex. Triplet states are in general hard to observe by electron resonance because of the

short relaxation times and the high anisotropy of the resonances. However, many such states have been observed, notably the excited triplet
states of aromatic hydrocarbons such as naphthalene, trapped in a particular orientation in a host crystal. Their decay time as observed by
electron resonance has been shown to be the same as the decay time
observed by phosphorescence studies.

7.10 Quadrupole effects

Fundamental nuclear theory proves that nuclei whose spin is greater than
one half will in general be non-spherical and hence have electric quadrupole moments. Three examples of a system with an electric quadrupole
moment are given in figure 7.15. A quadrupole moment will orient in a

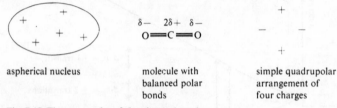

aspherical nucleus   molecule with   simple quadrupolar
                      balanced polar   arrangement of
                      bonds   four charges

Fig. **7.15** Three examples of electric quadrupoles

linear field gradient, just as a dipole orients in a linear field. Therefore
quadrupolar nuclei will be influenced by the direction in which the
molecule containing them is oriented at any one moment, unless they are
in a highly symmetrical environment such as that of the ^{14}N nucleus in
the ammonium ion. The effect of this force on the nucleus is to link its
motion with molecular motions and hence to provide a powerful new
spin-lattice relaxation mechanism. For this reason the nuclear resonances
of most nuclei of spin $I \geqslant 1$ are very broad except with highly symmetrical species, and are hence often either undetectable or not very
useful.

Occasionally the measurement of the nuclear resonance linewidth
provides useful information about the species involved. For example,
a ^{51}V ($I = 7/2$) magnetic resonance study of polyvanadates measured the
following linewidths: VO_4^{3-} (very narrow), $VO_3(OH)^{2-}$ (45 Hz), $V_2O_7^{4-}$
(50 Hz), $V_2O_6(OH)^{3-}$ (120 Hz), $V_3O_9^{3-}$ (120 Hz), $VO_2^+(H_2O)_n$ (300 Hz).
These linewidths support the theory that all the species except for
$VO_2^+(H_2O)_n$ are based on a VO_4 tetrahedron to which other groups can
be attached.[19] Any other arrangement of oxygens about the ^{51}V nucleus
would be expected to give rise to bigger linewidths.

In the absence of a magnetic field a quadrupolar nucleus orients
relative to any static field gradient it may experience. Transitions among
the different orientational states in solid samples may be induced in much
the same way as normal n.m.r. transitions, and the magnitude of the

[19] O. W. Howarth and R. E. Richards, *Journal of the Chemical Society*, 886, 1965.

transition energy is a measure of the field gradient. Experiments of this type are called nuclear quadrupole resonance experiments and are principally carried out in order to detect field gradients in crystalline samples.[20]

Suggestions for further reading

J. D. Roberts, *Nuclear Magnetic Resonance: Applications to Problems in Organic Chemistry*, McGraw-Hill, New York, 1959.
A. Carrington and A. D. McLachlan, *Introduction to Magnetic Resonance*, Harper and Row, New York, 1967.
R. M. Lynden-Bell and R. K. Harris, *Nuclear Magnetic Resonance Spectroscopy*, Nelson, London, 1969.
J. A. Pople, W. G. Schneider, and H. J. Bernstein, *High Resolution Nuclear Magnetic Resonance*, McGraw-Hill, New York, 1969.
P. B. Ayscough, *Electron Spin Resonance in Chemistry*, Methuen, London, 1967.

Problems

7.1 Calculate the Larmor frequency (in MHz) for the ^{51}V nucleus in a field of 1 T (10^4 gauss) from the data given in table 7.1. Also calculate its magnetogyric ratio.

7.2 An electron resonance experiment conducted at a frequency of 8.95 GHz detects a single resonance at a field of 0.3410 T (3410 gauss). What is the *g*-value of the free electron involved?

7.3 A free radical is dissolved in a solvent which has a detectable nuclear magnetic resonance. The spin-lattice relaxation of the solvent nuclei is dominated by the presence of the free radical in such a way that it is a fairly good approximation to say that every time an electron spin increases in energy due to spin lattice relaxation, there is a 0.01% chance that a nuclear spin will increase in energy, and every time an electron spin decreases in energy there is a 0.01% chance of a nuclear spin decreasing in energy.

    a Show that, if the electron resonance is saturated, the nuclear spin population difference will be increased until it reaches that which existed in the electron spin system before saturation. (This is known as dynamic nuclear polarization, or the Overhauser effect.)

    b Show that, within the approximation given, the nuclear resonance will appear saturated when the electron resonance is completely unsaturated.

    c Experimentally, the nuclear spin population difference is, in fact, quite unaffected by the unsaturated free electron spin system. Suggest a minor improvement to the initial approximation to include this fact and hence avoid the conclusion from **b**, while not seriously affecting the correct conclusion from **a**.

7.4 The width of the proton resonance in solid benzene is 1.8 gauss (1.8×10^{-4} T) below 90 K, but from 120 K to the melting point is only 0.9 gauss. Suggest a semi-quantitative reason for this.

7.5 The proton magnetic resonance spectrum of commercial hexadeuteroacetone shows a small resonance which may be resolved into a 1:2:3:2:1 pentuplet. Suggest a likely explanation.

7.6 Paraldehyde is a trimer of acetaldehyde. Its proton magnetic resonance

[20] Some interesting applications are described by M. Kubo and D. Nakamura in *Advances in Inorganic Chemistry and Radiochemistry*, **8**, 257, 1966.

spectrum shows a $1:3:3:1$ quartet at $\tau = 4.95$ and a $1:1$ doublet at $\tau = 8.60$, of three times the total area. What is the structure of paraldehyde?

7.7 Sketch the proton magnetic resonance spectra you would expect to obtain for

   a Methyl ethyl ketone

   b Tetramethylammonium ion

   c *n*-propane

   d A $50:50$ mixture of keto- and enol- forms of acetylacetone, with slow proton exchange.

Explain the features that you include and indicate the relative peak areas. In all cases ignore couplings through more than three bonds.

7.8 When pure naphthalene, dissolved in dichloromethane, is oxidized with antimony pentachloride, the product has an electron resonance spectrum which may be analysed into a regular nonet ($1:8:28:56:70:56:28:8:1$), each line of which is split by a smaller amount into a further nonet. The hyperfine splitting constants are each about one half of those predicted for the cation (naphthalene)$^{+}$. What possible structures could the oxidation product have?

7.9 The observed spectrum of the methyl radical in solution is a $1:3:3:1$ quartet of splitting -53.6 MHz. What structure does this suggest for the methyl radical? Estimate the splitting constant that you would observe if the radical were tetrahedral and the free electron were shared equally between the four sp^3 hybrid orbitals.

7.10 Type 1B diamonds differ stoichiometrically from pure diamonds in having about 0.1% of carbon atoms replaced by nitrogen atoms. The extra electron associated with each nitrogen atom goes into one of the four N—C bonds and lengthens it.

   What kind of electron resonance spectrum would you expect from a type 1B diamond at room temperature? How would it depend on the orientation of the diamond in the field? Reversible changes are observed when such diamonds are heated strongly in vacuo. Explain.

7.11 The intense ultraviolet photolysis at 26 °C of ethanol containing hydrogen peroxide gives rise to the electron spin resonance spectrum shown below. Identify the radical formed. What exchange processes are you assuming? How might you confirm your assignment?

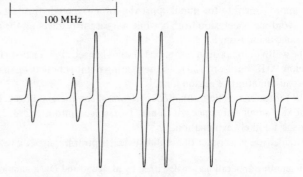

100 MHz

Problem **7.11**

7.12 The diagram shows the changes in the ^{51}V nuclear magnetic resonance spectrum of aqueous potassium orthovanadate as the solution is progressively acidified. The spectra are in first derivative form, and the species involved are labelled underneath at their chemical shift positions. Elucidate the equilibria involved.

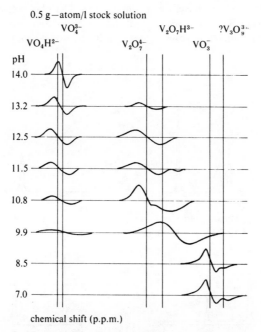

Problem **7.12** ^{51}V resonances in solutions of different pH

7.13 The figure illustrates time-dependent processes involving methanol. Spectrum (a) is of reasonably pure methanol in 50% solution in acetonitrile, and you are asked to analyse it, allowing for the fact that the slightly irregular peak heights arise because the chemical shift separation of the two types of proton has only about five times the magnitude of the spin-spin splitting at this field. Spectrum (b) is of the same methanol with only 5% of acetonitrile. In spectrum (c), about 10% of neutral water has been added. Finally, spectrum (d) shows the effect of adding about 30% of concentrated hydrochloric acid. Show how the spectral changes (a) → (d) may be explained in terms of the increasing exchange rate of the hydroxyl proton.

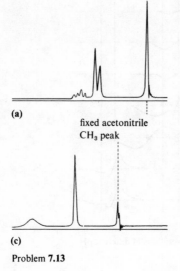

(a)

fixed acetonitrile
CH₃ peak

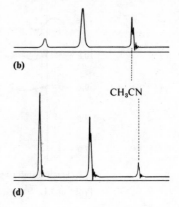

(b)

CH₃CN

(c)

(d)

Problem **7.13**

8 Pure rotational spectroscopy

8.1 Introduction

Within the limits of the Born-Oppenheimer approximation we have seen that linear molecules have two rotational degrees of freedom independent of other motions, and that other molecules have three degrees. The motions within these degrees of freedom are quantized, and therefore only certain total rotational energies are permitted. Changes of rotational energy may be directly induced by the action of electromagnetic radiation of microwave or far infrared frequency on molecules possessing a permanent electric dipole moment. The theory of this 'electric dipole rotational spectroscopy' is almost identical to the theory outlined for electron transitions in chapters 3 and 4; for x-polarized radiation one merely replaces $e\boldsymbol{E}_0 x$ with $\boldsymbol{E}_0 \mu_x$, where μ_x is the x-component of the permanent dipole moment. Changes may also be produced as a by-product of light scattering, even if the molecule does not possess a permanent dipole moment. These are discussed in a later section as rotational Raman spectroscopy.

The basic instrumentation of both the electric dipole and the Raman rotational spectroscopy of gases was described in chapter 5. The reasons why only gases give narrow absorption lines were explained in section 6.2(iv).[1] The interpretation of the results is far simpler with linear molecules than with non-linear, and therefore we discuss the former first.

8.2 Microwave rotational spectroscopy of rigid linear molecules

There is no simple classical way in which one can explain microwave rotational spectroscopy. Classical mechanics predicts absorption at all frequencies, and it is necessary to use quantum mechanics to predict discrete absorption lines. The classical absorption process is best understood by considering the microwave radiation to consist of two equal and opposite circularly polarized components (chapter 4). The molecular electric dipole will attempt to rotate parallel to the electric field of one of these components.

The quantum mechanical analysis of the rigid linear rotor is algebraically very similar to the determination of the angular dependence of

[1] For an account of more specialized microwave studies in condensed phases see D. H. Whiffen, *Quart. Rev.*, **4**, 131, 1950.

the hydrogen atom wavefunctions, and was discussed in section 2.3. The allowed rotational energies E are given by

$$E = \frac{\hbar^2 J(J+1)}{2I} \tag{8.1}$$

where J is integral and I is the moment of inertia of the molecule about any axis through the centre of gravity and perpendicular to the molecular axis. Electric dipole rotational transitions are governed by the selection rule $\Delta J = \pm 1$, which may either be deduced directly from the transition moment integral (section 4.7) or by following the argument of section 10.8 or of problem 10.8. The $J+1 \leftarrow J$ absorption is therefore at an energy

$$\Delta E = \frac{\hbar^2}{2I} \{(J+1)(J+2) - J(J+1)\} = \frac{\hbar^2(J+1)}{I} \tag{8.2}$$

and the entire absorption spectrum consists, on this simple model, of a series of lines evenly spaced in frequency by an amount $h/4\pi^2 I$. Half the amount of this frequency spacing is conventionally called the rotational constant B,[2] so that, for absorption,

$$\Delta v = B(J+1)(J+2) - BJ(J+1) \tag{8.3}$$

An experimentally idealized spectrum is shown in figure 8.1 for $^{16}O\,^{12}C\,^{32}S$.

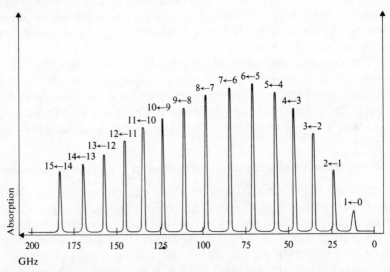

Fig. **8.1** Microwave rotational absorption spectrum of $^{16}O^{12}C^{32}S$

The main use of rotational spectroscopy is to extract molecular moments of inertia and hence bond lengths. This requires certain assump-

---

[2] Microwave spectroscopists, who measure frequencies in their spectra, define $B = h/8\pi^2 I$ as a frequency, and generally measure B in units of MHz $= 10^6\,s^{-1}$. Infrared spectroscopists, who measure wavenumbers, define $B = h/8\pi^2 cI$ as a wavenumber, and generally measure B in units of cm^{-1}.

tions. For a linear molecule whose moment of inertia is I one can assume rigidity, ignore the mass of the electrons, and thus deduce I from the definition

$$I \equiv \sum_n m_n r_n^2 \tag{8.4}$$

(m_n = nuclear masses, r_n = distance of these from molecular centre of gravity). With a diatomic molecule, equation 8.4 is sufficient to give the internuclear separation $r_1 + r_2$, because one already knows from the definition of centre of gravity that $m_1 r_1 = m_2 r_2$.

With a linear polyatomic molecule the complete set of r_n may be obtained either by knowing some lengths from other data, or by assuming some lengths or a particular symmetry, or by isotopic substitution. The latter, in its simple form, assumes that no bond length changes upon substitution of any of the masses m_n for isotopes of mass m_n'. Each substitution gives one more measurable I, and hence p substitutions are necessary for complete calculation of all the r_n of a linear $(p+2)$-atomic molecule. For example, the linear CO_2 molecule has only one unknown r_n, which in this symmetrical case is the C—O internuclear distance r_O. Hence $I = 2m_O r_O^2$. The linear OCS molecule, however, has three unknown r_n, namely the distances of each atom from the centre of gravity. One of these may be eliminated using the centre of gravity formula, but it is still necessary to make one isotopic substitution (e.g. ^{18}O for ^{16}O) to find all three r_n.

8.3   Non-rigid linear molecules

We have seen in chapter 2 that, even within the limits of the Born-Oppenheimer approximation, bonds are not rigid. Instead, the internuclear separation fluctuates very rapidly, so that the r_n of the previous section should more properly be interpreted as average distances. On this interpretation equation 8.4 becomes

$$I = \sum_n m_n \int r_n^2 f(r_n) \, d\tau \tag{8.5}$$

where $f(r_n)$ is the square of the normalized nuclear vibrational wavefunction of nucleus n, expressed in terms of r_n. The probability distribution $f(r_n)$ is usually almost symmetrical about the equilibrium r_n for the low vibrational states with which we are normally concerned. However when r_n is appreciably bigger than its equilibrium value it will contribute disproportionately to I. For this and other reasons, I and B will depend upon the extent of the nuclear vibrational wavefunction. This extent is greater for lighter nuclei, and therefore isotopic substitution by a heavier nucleus will increase I slightly less than predicted by equation 8.4. Hence the calculated r_n will be slightly too small for the heavier molecule. This is illustrated in table 8.1.

Table 8.1 Effect upon r_n of isotopic substitution

Molecule	$(r_H + r_{Cl})$ calc according to 8.4	Molecule	$(r_C + r_O)$ calc
$^1H^{35}Cl$	0.12837 nm	$^{12}C^{16}O$	0.11309 nm
$^2H^{35}Cl$	0.12813	$^{13}C^{16}O$	0.11308
$^3H^{35}Cl$	0.12800	$^{12}C^{18}O$	0.11308
$^1H^{37}Cl$	0.12837		

It can be seen from the table that the effect is most serious with hydrogen substitution. The relatively small effects in CO would be larger if the bond were weaker, and in calculations on larger molecules the errors due to multiple isotopic substitution will be cumulative unless these are corrected by equations such as 8.5.

A similar but larger change of I between different vibrational levels is discussed in section 9.9.

In order to obtain accurate values of I, a further important correction is almost always necessary. When a molecule rotates, its nuclei tend to pull apart under what is normally termed centrifugal force. Thus the actual molecular moment of inertia in states with non-zero J is greater than in the non-rotating state $J = 0$. The extent of the centrifugal distortion for a diatomic molecule is inversely proportional to the force constant for bond stretching, i.e. to the square of the vibrational stretching frequency v_{vib}, and a simple classical argument (problem 8.3) shows that the true rotational energies are given by the formula

$$E_{true} = BhJ(J+1) - DhJ^2(J+1)^2 \tag{8.6}$$

$B = h/8\pi^2 I_{J=0}$ as before, and D, the centrifugal distortion constant, equals $4B^3/v_{vib}^2$. Equation 8.6 shows that the percentage distortion of E is proportional to the rotational energy. It is a very reliable formula for all but the highest rotational energies, which shows that in this case the semi-classical explanation is almost as reliable as the full quantum-mechanical explanation involving the partial breakdown of the Born-Oppenheimer principle. Equation 8.6 is also valid for linear polyatomic molecules, although the simple relation $D = 4B^3/v_{vib}^2$ no longer holds for these.

8.4 Non-linear molecules

The discussion of the rotation of linear molecules is greatly simplified because there is only one relevant moment of inertia I. This moment of inertia is strictly defined in terms of Newton's second law. It is the constant of proportionality between any couple Fr applied about an axis through the centre of gravity and at right angles to the molecular axis, and the resulting angular acceleration $d^2\theta/dt^2$, so that

$$Fr = I\frac{d^2\theta}{dt^2} \tag{8.7}$$

However, in a two- or three-dimensional molecule we want to know the result of applying Fr about any axis we choose, through the centre of gravity. The resulting angular acceleration, and hence moment of inertia will depend on the axis chosen. Worse still, the angular acceleration that results from Fr will not always be about exactly the same axis direction as Fr. Theory shows that in general there are just three axis directions in space for which Fr is parallel to $d^2\theta/dt^2$. These are called the principal axes of the molecule, and the three resulting moments of inertia I_A, I_B, and I_C are called the principal moments of inertia. The theory also shows that, given I_A, I_B, I_C, and their directions it is possible to predict the rotational behaviour about any axis. The principal axes of small molecules are often obvious because of symmetry.

The fact that $d^2\theta/dt^2$ is not always parallel to Fr should be familiar to any amateur mechanic. For if one attempts to spin a wheel about an axis through its centre but not exactly at right angles to the wheel, the result is a juddering motion due to the non-parallel component of acceleration. Smooth spinning may only be achieved about the principal axis normal to the wheel (or, of course, the axes in its plane).

One particular simple type of non-linear molecule, whose rotational wave equation is readily solved, is the symmetric top, e.g. NH_3, CH_3Cl. Symmetric top molecules have the same symmetry as the wheel in the mechanical analogy above. For example, the C—Cl axis corresponds to the axis normal to the wheel, with moment of inertia I_A, and $I_B = I_C$, both being at $90°$ to I_A and to each other. CH_3Cl is in fact an example of a prolate (rugby-ball type) spherical top, where $I_A < I_B = I_C$ (see problem 8.7(a)). The other type is oblate (disc type, e.g. benzene) where $I_A = I_B < I_C$. Conventionally, $I_C > I_B > I_A$. One may consider a linear molecule to be an extreme example of a prolate symmetric top with $I_A = 0$, $I_B = I_C$.

The reason for the simplicity of the symmetric top is that its rotation about its unique principal axis has no effect upon, and is not affected by, its rotation about any axis at right angles to this. This is not true for molecules in general, as is illustrated in figure 8.2. Because $I_{B\,or\,C}$ in (a) can be about any axis normal to I_A its direction, once chosen, is not affected by molecular rotation about A. However in (b) the moment of inertia about any fixed spatial axis normal to I_C will vary in the event of rotation about C. This greatly complicates the theory of such rotations.

Because it is possible to separate rotation about the unique principal axis from other rotations in a symmetric top, one can treat this unique rotation in exactly the same way as the simple ring rotor discussed in problem 1.9. The angle of rotation about the unique axis is conventionally χ, the resulting wavefunctions $e^{+iK\chi}$, where K is a positive integer or zero, and the energies $E_{unique} = \hbar^2 K^2/2I_{unique}$. The angular momentum component about the unique axis is $\hbar K$, and we already know that the total angular momentum is $\hbar\sqrt{J(J+1)}$, where $J \geqslant K$. Therefore there must be an angular momentum of $\hbar\sqrt{J(J+1) - K^2}$ shared between the other two

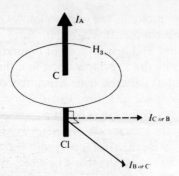

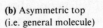

(a) Symmetric top, with CH_3 group drawn as a disc to illustrate its symmetry

(b) Asymmetric top (i.e. general molecule)

Fig. **8.2** Comparison of symmetric and asymmetric top molecules

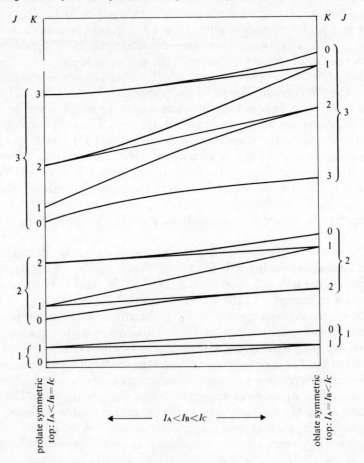

Fig. **8.3** Energy levels of the asymmetric top[3]

[3] From H. C. Allen and P. C. Cross, *Molecular Vib-Rotors*, Wiley, New York, 1963.

normal principal axes, so that

$$E_{\text{total}} = \frac{\hbar^2}{2I_{\text{other}}}(J(J+1)-K^2) + \frac{\hbar^2 K^2}{2I_{\text{unique}}} \tag{8.8}$$

The degeneracy of each level is the product of the J-degeneracy $(2J+1)$ and the K-degeneracy (2 except for $K = 0$ which is non-degenerate). In the special case of a 'spherical top' molecule, where $I_A = I_B = I_C$, K disappears from equation 8.8. Thus any one θ and ϕ wavefunction can be multiplied by any one of $2J+1$ different X wavefunctions giving the same overall energy, and so the total degeneracy is $(2J+1)^2$. Some of the degeneracy in both spherical and symmetric tops may be removed by centrifugal distortion.

The selection rules for the symmetric top are $\Delta J = \pm 1$, as with linear molecules, and $\Delta K = 0$. The χ motion cannot be affected by the electric field of radiation because it cannot involve any change in the direction of the molecular dipole moment. Therefore even with a symmetric top, $\Delta E = \hbar^2/I_{\text{other}}(J+1)$ for the $(J+1) \leftarrow J$ transition at any K.

The general wave equation for the asymmetric top, or general molecule, cannot be separated into θ, ϕ, and χ parts and hence can only be solved by complex numerical methods. The calculated lower levels for asymmetric tops are reproduced in figure 8.3.

A correctly analysed rotational spectrum yields up to three moments of inertia for the molecule. From these, with appropriate isotopic substitution, it is usually possible to calculate the bond angles and lengths with considerable accuracy, and also often to obtain centrifugal distortion constants. Symmetric and spherical tops and linear molecules only yield one moment of inertia, and planar molecules two, but their spectra then also reveal their symmetry. Microwave spectroscopy competes with

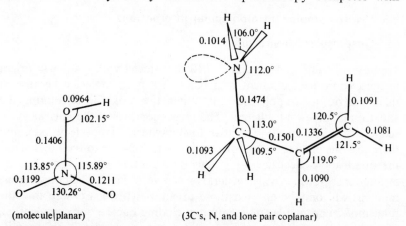

(molecule|planar)   (3C's, N, and lone pair coplanar)

Fig. 8.4 Precise molecular structure revealed by microwave spectroscopy[4] (bond lengths in nm)

[4] D. J. Millen and J. R. Morton, *J. Chem. Soc.*, 1523, 1960, G. Roussy, J. Demaison, I. Botskor, and H. D. Rudolph, *J. Mol. Spect.*, 38, 535, 1971.

X-ray diffraction as a prime means of obtaining precise molecular structures. Although X-rays can cope with much larger molecules, microwave spectra yield structures undistorted by crystalline intermolecular interactions. Because the microwave frequencies are often measurable to 1 part per million or better, bond lengths can be found in favourable cases to within 0.1 ppm $(= 10^{-3}\,\text{Å})$. This can reveal quite detailed distortions, as for the nitric acid and 3-aminoallene molecules in figure 8.4.

It has even proved possible to distinguish the *trans* (I) and *gauche* (II) isomers of 2 deuterio-ethylchloride,[5] for these two staggered rotamers are sufficiently long lived to give separate pure rotational absorptions.

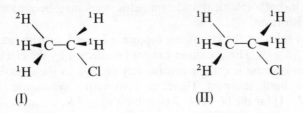

$$(I) \qquad\qquad\qquad\qquad (II)$$

Measurements have been made on molecules as large as quinuclidine (III) and azulene (IV).

$$(III) \qquad\qquad\qquad\qquad (IV)$$

although these are near the limits of analysability and vaporizability.

Microwave spectroscopy also shows some promise for the isotopic analysis of gases, because of the widely differing moments of inertia of isotopically substituted molecules.

8.5 More complex intramolecular interactions

(i) Quadrupolar fine structure

We have already seen in section 7.10 that nuclei with $I \geqslant 1$ will orient in the electric field gradient of a bond in a fixed molecule because of their electric quadrupole moment. When the molecule is rotating the situation is somewhat different. The precession axis of the nuclear gyroscope is fixed in direction relative to the external laboratory axes, and therefore this precession axis direction rotates relative to the bond. Thus the quadrupole–electric field gradient interaction energy becomes an average energy, which will in general be non-zero. The nuclear quadrupole moment will only be allowed those orientations which ensure that the total molecular angular momentum, including nuclear spin momentum, obeys the general quantum rules laid down in chapter 2. Because its allowed orientations are thus limited, it will have a limited number of allowed average quadrupole interaction energies, and therefore each

[5] R. H. Schwendemann and G. D. Jacobs, *J. Chem. Phys.*, **36**, 1245, 1962.

rotational level will be split by these energies. The spectrum splittings that result range typically up to 50 MHz, and are thus observable as fine structure under high resolution.[6]

(ii) Spin-rotational interaction

The electronic wavefunction of a molecule is normally calculated using the Born-Oppenheimer approximation of fixed nuclei. If the molecule is, in fact, rotating in one direction, the nuclei will always be slightly displaced from their assumed fixed positions. The electron cloud will therefore follow the molecular rotation slightly behind the nuclei. More precisely, the electron orbital momenta will no longer be exactly paired even in diamagnetic molecules. This results in a slight but detectable molecular magnetic moment, which is often measured in terms of a 'molecular g-value', by analogy with electron spin. If any nucleus has $I \geqslant \frac{1}{2}$, its spin magnetic moment will interact with this molecular magnetic moment and hence with the rotational motion, analogously to the quadrupole interaction described above. This interaction is observable in microwave spectra under the highest resolution and is known as 'spin-rotational interaction'. It is also an important additional cause of nuclear magnetic relaxation, especially in gases.

(iii) Hindered rotation

We have already seen that the *trans* and *gauche* isomers of 2-deuterio-ethylchloride give separate microwave spectra. The relative intensities of the absorptions at a known temperature give the relative energies of the rotamers. At higher temperatures the microwave spectrum becomes more complex, because in the presence of internal rotation[7] or torsional oscillation the rotational states of the molecule become substantially perturbed. The perturbation may be predicted by treating the internal rotation or torsion oscillation as a particle-in-a-ring motion with a suitable variation of potential energy with angle. Hence it is possible to discover not only the relative energies of 'fixed' rotamers but also the whole potential function for internal rotation. Typical three-fold internal rotation barriers thus observed have peak-to-trough energies of 4-15 kJ mol^{-1} (1-4 kcal/mole). In contrast, six-fold barriers are generally much lower in energy. Toluene, for example, has a six-fold barrier of merely 58.27 J mol^{-1} (13.94 cal/mole).

(iv) Inversion

Molecules such as ammonia and cyclobutane have two or more equivalent non-planar equilibrium configurations. Their inversion between these

---

[6] For further details see S. Walker and H. Straw, *Spectroscopy*, Chapman and Hall, London, vol. I, p. 135 et seq., 1961.

[7] This is in fact technically a vibrational motion, insofar as the Born-Oppenheimer separation retains any meaning.

configurations is, strictly, a molecular vibration. However, as will be seen from the potential functions in figure 8.5, some of the vibrational levels are so close that transitions between them may be detected in the micro-wave region. The split levels in (c) may be crudely considered as being split by the inversion frequency acting as an amplitude modulation of the bending mode. More properly, they represent separate nuclear states analogous to the electron 1s bonding and antibonding states of an H_2^+ molecule ion. Electric dipole transitions between these states are strongly allowed.

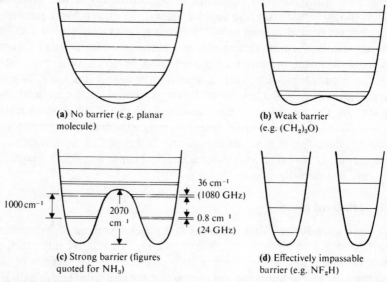

(a) No barrier (e.g. planar molecule)

(b) Weak barrier (e.g. $(CH_2)_3O$)

$1000\,cm^{-1}$

$2070\,cm^{-1}$

$36\,cm^{-1}$ (1080 GHz)

$0.8\,cm^{-1}$ (24 GHz)

(c) Strong barrier (figures quoted for NH_3)

(d) Effectively impassable barrier (e.g. NF_2H)

Fig. **8.5** Molecular inversion energy levels

Ammonia gas at room temperature has approximately 50% of molecules in each of the two lowest states. The states have slightly different total distributions of electric charge, and a beam of ammonia gas molecules may be physically separated into molecules of each state by a strong quadrupolar electric field. The excited state molecules, by themselves, will eventually lose their energy in a pulse, by a maser process analogous to the laser process described in section 10.13. In fact this was the first such pulse process to be discovered, in 1954.[8]

8.6 Stark effects

Mention has already been made in section 6.3 of the value of detecting a suitably modulated signal as a means of reducing unwanted baseline fluctuations. Such changes can be induced in the pure rotation micro-wave absorption spectra of gases by applying a fluctuating homogeneous electric field, usually with the aid of an insulated insert along the length of the waveguide sample cell.

[8] J. P. Gordon, H. J. Zeiger, and C. H. Townes, *Phys. Rev.*, **95**, 282, 1954, and **99**, 1264, 1955.

With a linear molecule, the electric field partially removes the $2J+1$ degeneracy of each level, because states with different $|m_J|$ correspond to different average angles between the molecular axis (i.e. dipole moment) and the applied field. Conversely two states with equal $|m_J|$ correspond to rotations at the same angle but in opposite senses. Unfortunately, all the simple linear rotor wavefunctions give the molecular axis an equal chance of being with or against the electric field, and so the Stark splitting only occurs to the extent that these wavefunctions are themselves distorted by the electric field E_0. In other words, first-order perturbation theory (appendix 2) gives zero Stark effect for a linear molecule, and one has to go to second-order,[9] to deduce a splitting proportional to E_0^2. The result of the full calculation is

Energy of distorted levels (except for $J = 0$)

$$= BhJ(J+1) - \frac{\mu^2 E_0^2}{2Bh} \cdot \left[\frac{3m_J^2 - J(J+1)}{J(J+1)(2J-1)(2J+3)} \right] \tag{8.9}$$

A typical shift at $100\,\mathrm{kV\,m^{-1}}$ is $4\,\mathrm{MHz}$ (OCS $2 \leftarrow 1$ transition). With symmetric tops there is in addition a much larger effect, proportional to E, provided that $K \neq 0$. In this case the molecular dipole moment is not parallel to the axis of rotation, and its direct interaction with the applied electric field does not average to zero but rather depends upon m_J. The first-order splittings for symmetric tops are given by the formula

$$\text{Energy of distorted levels} = \mu E_0 \cdot \frac{m_J K}{J(J+1)} \tag{8.10}$$

In both these cases an extra selection rule operates, namely $\Delta m_J = \pm 1$, or 0, depending on the microwave polarization. This is analogous to the corresponding rule for atomic spectra. Once again the dipole moment μ may be quite accurately extracted. Indeed in favourable cases the different values of μ in different vibrational states may be found.

8.7 Rotational Raman spectra

Raman spectra are discussed fairly fully in the next chapter and in appendix 1, because their main use until recently has been in the study of molecular vibrations. This section should therefore be read in conjunction with sections 9.1, 9.4, and 9.7. The recent development of laser Raman spectroscopy has made it easier to study Raman transitions due to change of rotational state only. Laser light is highly monochromatic, and therefore the Rayleigh line is sufficiently narrow not to overlap the pure rotational Raman lines near to it.

The rotational Raman effect is observable in principle with all molecules except spherical tops. The electric polarizability of all other molecules varies in general as the molecule rotates in the electric field of radiation, and this modulates the scattered frequency in the same way as

[9] Problem 4.4 suggests one general method for calculating second order perturbation effects.

a classical vibration as described by equation 9.4. Thus one observes Stokes and anti-Stokes lines very near to and evenly spaced about the central Rayleigh line, provided that the latter is sufficiently narrow. The modulation is at twice the frequency of rotation, because the polarizability goes through two maxima and minima during one complete rotation.

The quantum mechanical explanation of this Raman scattering is in terms of equation 9.6. The most salient point is, however, easily grasped. Scattering is a two photon process. Each time a photon is absorbed or emitted by a molecule, J must change by ± 1 (sections 9.9, 10.8). Therefore if J changes at all in rotational Raman scattering by linear molecules, it must change by 2; i.e. $\Delta J = \pm 2$. Thus the rotational Raman lines of linear molecules have approximately double the spacing of ordinary microwave rotational lines, their frequencies v_{Raman} being given by the formula

$$v_{\text{Raman}}(J+2 \leftarrow J) = v_{\text{vib}} + \frac{\hbar^2}{I} \cdot (2J+3) \tag{8.11}$$

Symmetric tops have the additional possibility of $\Delta J = \pm 1$, $\Delta K = \pm 1$.

Although rotational Raman spectroscopy has only recently acquired general usefulness, it has the advantage that it is applicable to all molecules other than those with spherical symmetry, and does not require the presence of a molecular electric dipole moment.

8.8 Intensities

The intensity of an allowed pure rotational absorption is governed by the product of several separate factors, namely:

a the square of the permanent dipole moment, i.e. the square of the transition moment;

b the Boltzmann fractional population of each lower m_J sublevel. For a given J, relative to $J = 0$, this is $e^{-\hbar^2 J(J+1)/2IkT}$;

c the number of allowed transitions between the m_J sublevels of rotational state J and of state $(J+1)$. This is fixed by the $\Delta m_J = \pm 1$ or 0 selection rule, and is simply $3(2J+1)$. In other words there are three possible transitions from each m_J sublevel of state J;

d the factor ω_k in equation 4.19;

e reduction of absorption signal due to stimulated emission, as described in section 5.8.

In rotational Raman spectra **a** is replaced by expression 9.6. Also a further factor can arise in such spectra if the molecule has symmetry. For example, the pure rotational Raman spectrum of N_2, shown in figure 8.6 shows an intensity alternation between successive lines.

The spectrum appears to consist of two separate superimposed spectra, one for transitions involving states with odd J and one of twice the intensity for states with even J. Each sub-spectrum follows the con-

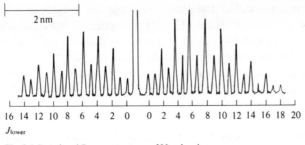

Fig. **8.6** Rotational Raman spectrum of N_2, showing alternating intensities[10]

siderations above, rising first as $(2J+1)$ increases and then falling because of the Boltzmann factor. The existence of two sub-spectra is one of the most revealing pieces of evidence for the Pauli exclusion principle in its general form. In developing his principle Pauli first used some very abstract physics to prove that there is an intrinsic difference between fundamental particles with zero spin and particles with a spin of one half. He showed that if you take the combined wavefunction ψ for two identical particles with zero spin, and then exchange the particles, the resulting wavefunction is also ψ. Conversely, if the exchanged particles have spins of one half, the resulting wavefunction is $-\psi$. In both these cases the particle density before and after exchange is $\psi^*\psi$, and is therefore unchanged, as must be the case with identical particles. The consequences of this for groups of electrons are discussed in section 10.3, where it is shown that, as a result, no two electrons may occupy exactly the same quantum state.

The consequences for nuclei are slightly more general. A nucleus consists of protons (spin $= \frac{1}{2}$), neutrons (spin $= \frac{1}{2}$) and various other less readily identifiable fundamental particles with zero spin. If two identical nuclei are exchanged, this amounts to the separate exchange of each of their constituent fundamental particles. The exchange of the spin zero particles does not affect ψ. However, ψ changes sign once for each proton or neutron that is exchanged. Therefore, a nucleus with an even total number of protons + neutrons will behave, on total exchange, like a spin zero particle. Such nuclei will always have integral or zero total spin, because there is no way in which an even number of half-integral spins may combine to give a half-integral spin. They belong to the group of particles called 'bosons'. Conversely, a nucleus with an odd number of protons + neutrons will always have half-integral spin; there is no way in which the spin of the final nucleon can be paired or doubled. Particles with half-integral spin are called 'fermions', and obey the same exclusion principle as do electrons, for the same reason. Bosons, however, may occupy a quantum state in any number; this helps to explain the peculiar properties of liquid ^{4}He, such as superfluidity.

[10] Reproduced from S. S. Porto, L. E. Cheeseman, A. Weber, and J. J. Barrett, *J. Opt. Soc. America,* **57**, 19, 1967.

The consequences of the Pauli principle for molecules are best explained by an example. If the ^{1}H—^{1}H molecule is rotated about 180° so that $\phi \to (\pi + \phi)$ and $\theta \to (\pi - \theta)$, this is equivalent to exchanging the two nuclei. The internuclear separation, R, remains of course unchanged. The total nuclear wavefunction must therefore change sign, because protons are fermions. The Born-Oppenheimer argument[11] tells us that:

$$\psi^{\text{nuclear}} = \psi(R)\psi(\theta, \phi)\psi(\text{spin})$$

Now $\psi(R)$, like R, is unaffected by internuclear exchange. The rotational wavefunction $\psi(\theta, \phi)$ changes to $-\psi(\theta, \phi)$ if J is odd, and is not changed if J is even, when $\theta \to (\pi - \theta)$ and $\phi \to (\pi + \phi)$. This may be checked using the $J = 0$, 1, and 2 wavefunctions of section 2.2.

The total spin wavefunction $\psi(\text{spin})$ is either one from a triplet state, in which case it is symmetric, or a singlet state, in which case it is antisymmetric with respect to nuclear exchange. The arguments for this are described in section 7.9(i). These are effectively separate states, so that molecular hydrogen may be considered at any one moment to be a mixture of four different species, three with symmetric spin wavefunctions (and some interchangeability) and one with an antisymmetric spin wavefunction. Because ψ^{nuclear} must change sign on exchange, this latter, called para-hydrogen, can only exist in rotational states with even-valued J. Similarly the former, ortho-hydrogen, can only exist in states with odd J. Hence we have explained the apparently superimposed spectra, and the $3:1$ intensity ratio. In ^{16}O—^{16}O $(I = 0)$ the effect is even more dramatic, because states with odd J are completely ruled out. In ^{14}N $(I = 1)$ the ratio is 6 (even J): 3 (odd J).

The nuclear spin restrictions upon rotational state occupancy are also important in calculations of specific heats. This is particularly true for hydrogen, where the $J = 1$ level is only significantly occupied at about 100 K. In general, molecules with an inversion centre have about half the rotational heat capacity of similar unsymmetrical molecules.

When several nuclei, or non-linear molecules, are involved the same principles apply, but their application is more complex. The inversion operation must be considered for all the nuclei in the molecule. Thus the inversion of ^{1}H—^{12}C—^{12}C—^{1}H is like that of ^{1}H—^{1}H, because the additional exchange of the ^{12}Cs does not change the sign of ψ. Conversely the inversion of ^{1}H—^{13}C—^{13}C—^{1}H changes ψ twice in sign, so that the whole nuclear wavefunction must be symmetric with respect to particle exchange. Nuclear spin wavefunction symmetry must also be taken into account in studies of internal rotations, e.g. of —C—^{1}H$_3$ groups. In this particular case the spins group to form a quartet (symmetric ψ (spin)) and a doublet (asymmetric) and so this $2:1$ weighting applies to the internal rotational levels with odd and even quantum numbers respectively.

[11] Even though this argument is only approximate, the correction terms needed to construct the true wavefunction will not change its symmetry from that of the approximate one.

Suggestions for further reading

D. J. E. Ingram, *Spectroscopy at Radio and Microwave Frequencies*, Butterworths, London, 1967 (2nd edn).

J. E. Wollrab, *Rotational Spectra and Molecular Structure*, Academic Press, New York, 1967.

C. H. Townes and A. L. Schawlow, *Microwave Spectroscopy*, McGraw-Hill, New York, 1955.

Problems

8.1 **a** The following absorption frequencies were observed for the linear molecule ^{1}H—^{12}C≡^{31}P

J	*frequency* (MHz)
$1 \leftarrow 0$	39 951.98
$2 \leftarrow 1$	79 903.31
$3 \leftarrow 2$	119 853.67
$4 \leftarrow 3$	159 802.56
$5 \leftarrow 4$	199 749.42
$6 \leftarrow 5$	239 693.82
$7 \leftarrow 6$	279 634.66
$8 \leftarrow 7$	319 573.03
$9 \leftarrow 8$	359 506.14

Use the first two values to obtain B. Then calculate D for each transition paired against $1 \leftarrow 0$, and show that an approximately constant value is obtained.

b The corresponding $1 \leftarrow 0$ value for ^{2}H—^{12}C≡^{31}P is 33 968.73 MHz. Use this fact to calculate the internuclear distances, assuming fixed nuclei and no isotope effects. (Mass of ^{1}H = 1.6722, ^{2}H = 3.395, ^{12}C = 20.227, ^{31}P = 52.241, all × 10^{-24} g.)

8.2 Calculations on isotopically different pairs of OCS molecules reveal slight variations in bond length, as shown in the table below.

pair	C=O	C=S (nm)
$^{16}O^{12}C^{32}S$ and $^{16}O^{12}C^{34}S$	0.1165	0.1558
$^{16}O^{12}C^{32}S$ and $^{16}O^{13}C^{32}S$	0.1163	0.1559
$^{16}O^{12}C^{32}S$ and $^{18}O^{12}C^{32}S$	0.1155	0.1565

Explain these discrepancies qualitatively.

8.3 Prove the relation $D = 4B^{3}/v_{vib}^{2}$ for a linear molecule, where D and B are as in equation 8.6 and v_{vib} is the vibrational frequency of the molecule. (*Hint*: convert v_{vib} to k, the force constant in the equation force = k × extension. Then calculate the effect of the extension upon I for a centrifugal force due to a rotational angular momentum of $\hbar\sqrt{J(J+1)}$. Hence find E_{rot}, to which must be added the potential energy of bond stretching.)

8.4 It is apparent from figure 8.6 that the $8 \leftarrow 6$ Raman transition in N_{2} is the strongest at the temperature used. Calculate this temperature, given that B for ^{14}N—^{14}N is 59 GHz, and using the formula

relative intensity of $(J+2) \leftarrow J = (2J+1)e^{-BhJ(J+1)/kT}$
(even J or odd J but not both)   [answer: ca. 240 K]

8.5 Classify the following molecules by their rotation types: CO_2, O_3, NH_3, ferrocene, C_2H_2, chair cyclohexane, $Ni(CO)_4$. Indicate whether the symmetric tops are prolate or oblate, and whether or not the molecule has a centre of inversion.

8.6 Assuming that two moments of inertia could be found for HNO_3 (and for isotopically substituted nitric acid) and that it was known to be planar, what is the minimum number of isotopic substitutions necessary to obtain the data in figure 8.4?

8.7 **a** Show that CH_3Cl has two equal moments of inertia about two perpendicular axes at 90° to the C—Cl bond direction. (*Hint*: take one axis along a CH bond direction in the plane-projected molecule.)

b Show similarly that the benzene molecule has two equal moments of inertia in its plane. Also show that it has exactly double this moment of inertia about an axis through its centre and perpendicular to its plane.

8.8 **a** Using Pythagoras' theorem, prove that for any planar body the sum of the moments of inertia about any two perpendicular axes in the plane equals the moment of inertia about the axis perpendicular to the plane axis and through the crossing point of the in-plane axes.

b The moments of inertia of most planar molecules almost obey this theoretical relationship. However, there is almost always a small discrepancy, known as the 'inertia defect'. Suggest an explanation for this.

8.9 **a** Explain carefully why alternating peak intensities are never observed in microwave rotational spectra of linear molecules.

b Suggest an experimental means of proving that the rotational degeneracy of the spherical top is $(2J+1)^2$.

8.10 Explain why the ^{16}O—^{16}O molecule may never exist in a rotational state with odd J.

8.11 Calculate the relative rotational Raman intensities for odd J and even J for the molecules ^2H—^2H, ^1H—^{13}C—^{13}C—^1H. (^2H has $I = 1$ and ^{13}C has $I = \frac{1}{2}$.)

8.12 The HSSH molecule is an asymmetric top in which I_B is very nearly equal to I_C. Its microwave spectrum is very like that of a symmetric top, except that one also observes transitions with $\Delta K = \pm 1$. Explain this.

9 Infrared and vibrational Raman spectroscopy

9.1 Introduction

The nuclei in a molecule are in constant motion relative to each other. We saw in chapter 2 how any complex molecular motion may be analysed (at least approximately) as a superimposition of 'normal' modes of motion, and how those modes of nuclear motion not involving translation or rotation were the normal vibrational modes of the molecule.

Also, in chapter 1, we saw how Planck was forced through his studies of black body radiation to deduce that the energy of any vibrational mode could only be taken up or given out in discrete quanta of energy $h\upsilon_{vib}$, where υ_{vib} was the frequency of oscillation of the vibrational mode involved. This deduction was supported by subsequent detailed solution of the Schrödinger equation, and is nowadays verified most plainly by the existence of two branches of spectroscopy, infrared and Raman.

Ordinary infrared spectroscopy should more strictly be described as infrared absorption spectroscopy, as it measures the absorption of infrared radiation of frequency υ as the molecules of the sample gain vibrational energy in quanta of $h\upsilon$. Raman spectroscopy, however, measures emission and absorption of vibrational quanta in one spectrum. When visible light is scattered (see chapter 4) from a molecular sample, a small percentage of the scattered light is reduced in frequency by amounts υ_{vib}. This effect was first observed by C. V. Raman, an Indian spectroscopist, in 1928. Also a somewhat smaller percentage of the scattered light is increased in frequency by υ_{vib}. Thus the spectrum of the scattered light consists of a very strong peak at the original incident frequency υ_0 flanked by equally spaced small peaks at frequencies $\upsilon_0 \pm \upsilon_{vib}$. The peaks at $\upsilon_0 - \upsilon_{vib}$ are called Stokes lines, and the weaker ones at $\upsilon_0 + \upsilon_{vib}$ anti-Stokes lines. (Normally only the former are recorded.)

The practical details of infrared and Raman spectroscopy are included in chapters 5 and 6. We now proceed to study the vibrational quantum and selection rules involved, first for the idealized case of the simple harmonic oscillator, and then for anharmonic oscillators and linked oscillators, as in real molecules.

9.2 Simple harmonic oscillator

The simplest model of a molecular oscillator consists of two nuclei linked by a chemical bond, such that the force between the nuclei obeys

Hooke's law—i.e. is proportional to the compression (or extension) of the bond from its equilibrium length. This model is quite a good approximation to typical diatomic molecules such as N_2 and HCl provided that they are not compressed or extended too far from their equilibrium length. The reason is apparent from figure 9.1, which plots the energy of a typical diatomic molecule against the internuclear separation, R. The bottom part very nearly coincides with the corresponding parabolic energy vs. separation curve (dotted) for a simple harmonic oscillator (see problem 9.1).

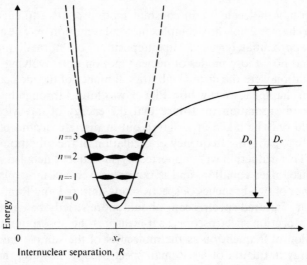

Fig. **9.1**

Figure 9.1 also shows some of the energy levels predicted when Schrödinger's equation is solved for the simple harmonic oscillator (i.e. for nuclear particles in the 'field' of a chemical bond that obeys Hooke's law). The predicted probability densities for the particles, as functions of R, are also indicated by the varying line density. The wavefunction corresponding to one of these probability densities is drawn in figure 9.2, together with the probability density predicted by classical mechanics. It is important to notice the differences between these two. Classical mechanics predicts that, for a given energy of vibration, the limits of R are given precisely by the points of intersection of the parabola of figure 9.1 with the horizontal line representing the particular total vibrational energy under consideration. It also predicts that the nuclei are most likely to be at their maximum or minimum separation (where they are moving most slowly) and least likely to be at their equilibrium separation (where they are moving most rapidly).

For large vibrational energies the classical description is a fair approximation to the quantum-mechanical one. However, at lower energies major differences are apparent. First of all, a real diatomic molecule (i.e.

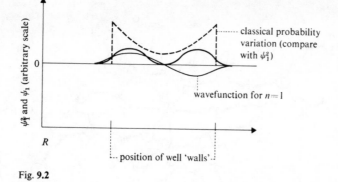

Fig. **9.2**

one obeying the laws of quantum mechanics) is not confined within certain rigid maxima and minima of internuclear separation. Indeed, it is meaningless to talk of the internuclear separation as a precisely measurable quantity for a given energy. One can only talk about the probability densities for the nuclei, and one finds that this probability density is not zero outside the classical limits of R. There is a small but finite chance of finding the oscillator with a very large or a very small internuclear separation. This explains one example of a phenomenon known as 'tunnelling', of which the best-known example is radioactivity; the probability of fission of an unstable isotope is directly proportional to the tiny but non-zero probability that some of the nuclear particles have of being outside the nuclear potential well.

The reality of the spread-out nuclear wavefunction is nowadays evident from very high resolution, low temperature X-ray studies. These can be of sufficient resolution to reveal the extent and shape of the nuclear probability cloud, and to confirm the further prediction of quantum mechanics that a wide potential well (e.g. weak restoring force) for a particular nucleus in a particular direction leads to a diffuse nuclear wavefunction along that direction. Figure 9.3 provides an example of this.

The second major difference between the predictions of classical vibrator theory and those of quantum theory is, of course, that the latter predicts that only certain vibrational energies are permitted. For the harmonic oscillator, these energies are in fact $(n+\frac{1}{2})h\upsilon$,[1] where n is an integer and $\upsilon = \upsilon_{vib}$ = the classically calculated vibrational frequency. The minimum possible value for n is zero, which means that even at the absolute zero of temperature a molecule has some vibrational energy, normally called 'zero-point energy'. This surprising conclusion is illustrated by very accurate calculations and observations of the bond energies of H_2, HD, and D_2. Because the deuterium nucleus has a mass number of 2, the zero-point energy of D_2 is $1/\sqrt{2}$ times that of H_2 and hence the thermodynamic bond energy is slightly greater (see table 9.1).

[1] This is a well-known result, and its proof may be found, for example, in L. Pauling and E. B. Wilson, *Introduction to Quantum Mechanics*, McGraw-Hill, New York, 1935.

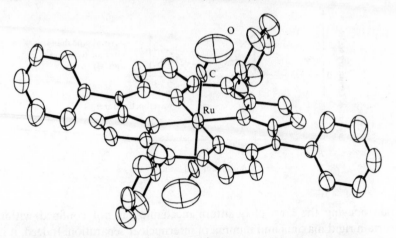

Fig. **9.3** X-ray structure of dicarbonyl-*meso*-tetraphenylporphinato ruthenium II (excepting H atoms). Each ellipsoid shows the region within which the nucleus has a 50% probability. The size of the ellipsoids is mainly due to zero-point motion plus bulk crystal vibrations. The lack of angular rigidity of the CO ligands is striking. (D. Cullen, E. Meyer jun., T. S. Srivastava, and M. Tsutsui, *Chem. Comm.*, 1972, 584.)

The changes in the bulk physical properties of compounds upon isotopic substitution are largely explained by zero-point energy changes.

Table **9.1**[2] D_0 (i.e. ΔH for dissociation) and zero-point energy of H_2, HD and D_2

	Experimental (and theoretical) D_0 kJ mol^{-1}	Calculated z.p.e. kJ mol^{-1}	$D_e (= D_0 + z.p.e.)$
H_2	432.04	26.07	458.11
HD	435.45	22.62	458.07
D_2	439.59	18.50	458.09

9.3   Mechanism of infrared absorption

The absorption of radiation by a dipole system was discussed in chapters 3 and 4, with particular reference to an electron-nucleus dipole. The requirement for interaction between the dipole system and the radiation was that the overall dipole moment changed as the electron and nucleus moved relative to each other. In this way the classical dipole was able to counteract the electric field of the radiation by oscillating at the radiation frequency, and the quantum mechanical dipole was given a finite chance of gaining or losing energy so as to reach a different energy state.

Most molecules change their overall dipole moment as they vibrate. The hydrogen chloride molecule, for example, acquires an increasing dipole moment as the H—Cl bond stretches. Radiation interacts with such vibrating dipole moments. Hence infrared spectra. However, there are a few cases where a vibration does not change the dipole moment, in

[2] Data from G. Herzberg, *J. Mol. Spectroscopy*, **33**, 147, 1970, and references contained therein.

particular the symmetrical stretching vibrations of centrosymmetric molecules such as nitrogen and carbon dioxide. Such vibrations do not interact directly with radiation and hence do not directly give rise to infrared absorptions. The formal requirement that the vibration must change the dipole moment of the molecule in order to be infrared active is known as a 'gross selection rule'.

In addition to this, there is an ordinary quantum mechanical selection rule for the simple harmonic oscillator that $\Delta n = \pm 1$. This means that idealized molecules can only gain or lose one quantum of vibrational energy at a time. A proof of this is given in appendix 1. The selection rule is substantially confirmed by experiment, but requires some modification if the oscillator is not strictly simple harmonic.

When the selection rule $\Delta n = \pm 1$ is combined with the energy level equation

$$\text{Energy of level } n = (n + \tfrac{1}{2})h\upsilon_{\text{vib}} \tag{9.1}$$

it becomes apparent why, as proposed originally by Planck, radiation can only be absorbed (or emitted) at frequency υ_{vib}. This is precisely the classical conclusion, reached in a different way. The quantum mechanical derivation should be understood as the 'correct one', because strictly speaking the vibration frequency is not directly measureable. It is merely a convenient way of describing a particular parabolic potential energy well. When the potential well is not parabolic the predictions of classical and of quantum mechanics differ and experiment confirms the latter rather than the former.

9.4   Mechanism of Raman scattering

Raman scattering is much more complex than ordinary infrared absorption, and the full quantum mechanical theory is still being worked out. Once again, the classical theory is able to explain the existence of the effect, whilst quantum theory is needed to predict selection rules and intensities correctly.

Classical theory

Let us illustrate the classical theory by considering a diatomic molecule such as hydrogen. Like the atoms considered previously, this has a polarizability α_z along the bond axis such that if it is put into an electric field E_z along that direction then a dipole moment μ_z results. However, unlike atoms, the molecular polarizability depends on the internuclear separation (and also on the orientation of the molecule with respect to the electric field). The polarizability contribution of any electron in a particular atomic or molecular orbital is approximately proportional to the volume of the orbital, and therefore as the molecule expands and contracts during its (classical) vibration its polarizability increases and decreases. If the vibration frequency is v_{vib} then to a first approximation its bond length $R = R_e$ (the equilibrium length) $+ r \cos 2\pi v_{vib} t$, where r is the vibration amplitude, and

$$\alpha_z = \alpha_{0z} + r \frac{d\alpha}{dR} \cos 2\pi v_{vib} t \qquad (9.2)$$

In other words the polarizability fluctuates sinusoidally about a mean of α_{0z}, with a fluctuation proportional to the amplitude of vibration and to the rate of change of polarizability with bond length around R_e.

Now let us suppose that the electric field E_z is due to radiation of frequency v_0. We may then write

$$E_z = E_{0z} \cos 2\pi v t \qquad (9.3)$$

and by combining this with equation 9.2 we get

$$\mu_z = E_z \alpha_z = E_{0z} \alpha_{0z} \cos 2\pi v_0 t + E_{0z} r \frac{d\alpha}{dR} \cos 2\pi v_{vib} t . \cos 2\pi v_0 t \qquad (9.4)$$

The first term was discussed in chapter 4, and gives rise (on the classical model) to Rayleigh scattering. The second term of equation 9.4 may be rewritten:

$$\tfrac{1}{2} E_{0z} r \frac{d\alpha}{dR} \cos 2\pi(v_0 + v_{vib})t + \tfrac{1}{2} E_{0z} r \frac{d\alpha}{dR} \cos 2\pi(v_0 - v_{vib})t \qquad (9.5)$$

(because $2 \cos A \cos B = \cos(A+B) + \cos(A-B)$).

Thus the classical theory predicts that, in addition to Rayleigh scattering, there will be weaker lines at frequencies $v_0 \pm v_{vib}$, and it identifies these with the observed anti-Stokes and Stokes lines.

Quantum theory

The quantum mechanical theory of light scattering is quite different from the classical theory, and is not easily understood. This section is therefore best omitted at a first reading. The theory involves second-order time-dependent perturbation theory, which is an extension of the first-order theory outlined in chapter 4. In that chapter, equation 4.12 was transformed into equation 4.13 by making the approximation that all the a_k were negligible in comparison with the time-dependent coefficient of the ground state. With this approximation it was possible to calculate the (small) a_k. One can make a better approximation (accurate to 'second-order') by using these approximate values for a_k instead of setting all the upper state a_j's equal to zero. After a long and tedious calculation one derives[3] the result that there is a second indirect probability of transition

---

[3] This result is proved in many advanced texts, e.g. H. F. Hameka, *Advanced Quantum Chemistry*, Addison Wesley, Reading, Mass. There are some additional smaller terms as well.

between the ground state G and any excited state j primarily proportional to the square of

$$\sum_k \frac{\int \psi_j^* eE_0 r\psi_k \, d\tau \cdot \int \psi_k^* eE_0 r\psi_G \, d\tau}{\omega_{jk} + \omega} \qquad (9.6)$$

where $-\omega_{jk}$ is the frequency corresponding to the energy jump from state j to any other state k (usually k is a state of much higher energy, and j is near to G, the ground state). If j is simply the first excited vibrational state, the above probability, multiplied by the fraction of vibrators in the ground state, governs the intensity of the Stokes Raman lines.

By reversing j and G, the reverse process (i.e. the appearance of anti-Stokes lines) is also predicted, although with lower intensity because of the lower initial population of the excited state j from which transitions occur. Thus the quantum theory is able to explain the observed differences in intensity between the Stokes and the anti-Stokes lines, whereas the classical theory predicted equal intensities.

The chief difficulty with expression 9.6 is that because its derivation is lengthy, its physical meaning is obscure. However, there are various rather crude ways in which it may be analysed. The most useful is to consider it as describing two linked steps, namely the absorption of a photon taking the system from stage G to state k followed immediately by the emission of a photon taking the system back to state j. This rationalizes the presence of the product of the two transition moment integrals.

The indirect transition probability described by expression 9.6 carries hidden within it the selection rule $\Delta n = \pm 1$, where n is the vibrational quantum number between state G and state j and where states G and j are assumed to have identical electronic wavefunctions, as is normally the case. As with the similar infrared selection rule, this rule is due to the detailed properties of the nuclear vibrational wavefunctions. It is partially derived in appendix 1.

A further gross Raman selection rule, applicable to polyatomic molecules, is considered in section 9.7.

9.5 Applications of infrared and Raman measurements on simple oscillators

A diatomic molecule or ion normally gives rise to a single infrared or Raman absorption line. The exceptions to this are with high temperatures (hot bands) and with low pressure gases (rotational fine structure), which will be considered later. We have seen how the frequency of the absorption line is related to the frequency v_{vib} of the vibration. For the simple harmonic oscillator the restoring force F is proportional to the displacement $(R - R_e)$, i.e.

$$F = k(R - R_e) \qquad (9.7)$$

and k is known as the force constant of the bond. Elementary mechanics shows that for a diatomic molecule with two nuclei of masses M_1 and M_2 linked by electrons of negligible mass, the relation between k and v_{vib} is

$$(2\pi v_{vib})^2 = \frac{k(M_1+M_2)}{M_1 M_2} \tag{9.8}$$

The force constant of a bond does not depend directly on the masses of the nuclei it links. Therefore, we would expect the relative vibration frequencies of isotopically substituted diatomic molecules to be simply related to their isotopic masses. For example, the mass number of the proton is 1.00782, and that of the deuterium nucleus is 2.01410. Hence we would expect from equation 9.8 that their vibration frequencies would be in the ratio $H_2 : D_2 = 1 : 0.7074$. The experimental ratio is in fact $1 : 0.7095$.

The theoretical calculation of force constants is not easy for heavy molecules. For example, the best calculated value for CO is in error by about 10%. However, the best value for H_2 is within 0.02% of the experimental value. Despite the theoretical limitations, force constants remain a useful approximate guide to the strengths of comparable bonds. This is because the compressibility of a bond is directly proportional to its total energy at its equilibrium length, as well as being proportional to other factors such as the general extent of the atomic orbitals involved in bonding. Table 9.2 shows some force constants and corresponding bond strengths for diatomics.

Table 9.2

	H_2	N_2	O_2	Cl_2	KBr	ZnH
Force constant relative to H_2	1	4.00	2.05	0.57	0.14	0.26
Bond strength relative to H_2	1	2.18	1.13	0.55	0.88	0.19

(The high force constant of N_2 may be due to the strong separation dependence of π-bond overlap, and the low force constant of KBr may be due to the weak, square law dependence of the largely ionic bonding.)

The single infrared or Raman absorption line arising from a diatomic species is a moderately sensitive way of detecting that species. For example, the presence of the Hg_2^{2+} cation in aqueous solutions of mercurous salts was strongly confirmed by the detection of the Hg—Hg stretching mode in the Raman spectrum.

9.6   Polyatomic molecules

We have seen in chapter 2 that the $3n$ degrees of freedom of a non-linear n-atom molecule may be analysed into three translational modes of motion, which account for all the motion of the centre of gravity, three rotational modes, which account for all possible overall rotation of the molecule, and $3n - 6$ vibrational (or internal rotation) modes. Similarly a linear n-atom molecule has three translational, two rotational and $3n - 5$ vibrational modes. These modes of motion are for the nuclei, and exclude electron motions and nuclear and electron spin. We also saw that the separation of the molecular motion into these modes was only made possible by the Born-Oppenheimer approximation.

Each normal mode of vibration has its characteristic 'fundamental' frequency, and this frequency will almost always appear as a strong absorption in either the infrared spectrum, the Raman spectrum, or both, according to the gross selection rules explained in section 9.7. Other frequencies may also appear in the infrared spectrum, and the reasons for this are explained in section 9.8. The observed frequencies in the infrared and Raman spectra of a substance may often be correlated with the expected modes of vibration (when these are known) by empirical arguments about the likely fundamental frequencies of the modes, and by observation of related isotopically substituted species.

The example of CO_2 was discussed. A slightly more complex example, which illustrates the general principles of elementary vibrational mode analysis, is that of acetylene. Acetylene is a linear molecule which must therefore have $3 \times 4 - 5 = 7$ vibrational modes. The problem of identifying these is the problem of dividing them into distinguishable types. One may first of all distinguish between stretching and bending modes. The stretching modes of a linear molecule do not involve any changes of bond angle. Therefore, as there are three bonds in C_2H_2 we may expect to find three stretching modes. The stretching motions may be further analysed, in C_2H_2, into the symmetric and antisymmetric combinations of the individual C—H stretching vibrations, plus the C—C stretch (with the C—H distance little affected). These make up the modes known as v_1, v_3, and v_2 respectively in figure 9.4. The justification for taking symmetric and antisymmetric combinations of the individual C—H stretches is that

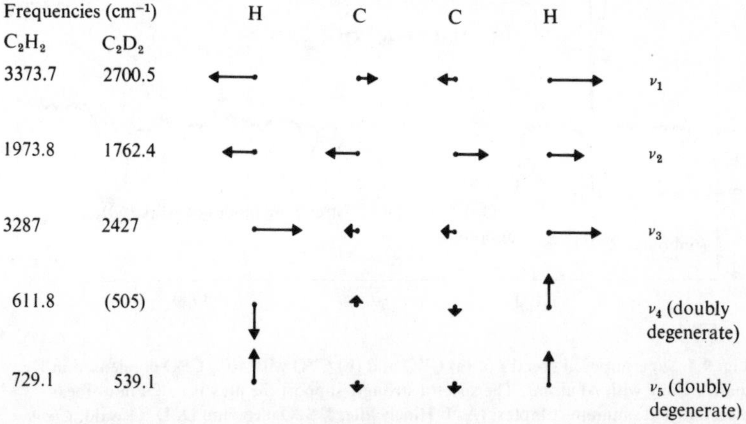

Frequencies (cm⁻¹)		H	C	C	H	
C_2H_2	C_2D_2					
3373.7	2700.5	←	→	←	→	v_1
1973.8	1762.4	←	←	→	→	v_2
3287	2427	→	←	←	→	v_3
611.8	(505)	↓	↑	↓	↑	v_4 (doubly degenerate)
729.1	539.1	↑	↓	↓	↑	v_5 (doubly degenerate)

Fig. **9.4** Normal vibrational modes of the acetylene molecule the uncombined stretching vibrations are of the same frequency, and hence combine to give modes in which all atoms move with the same frequency—i.e. normal modes.

This leaves four bending modes, corresponding to changes in the two HCC bond angles in each of the two normal planes containing the undistorted C_2H_2 molecule. Again, one may analyse these into symmetric (v_5) and antisymmetric (v_4) simultaneous changes of HCC bond angle. Each of these modes is doubly degenerate, like the bending mode of CO_2. It will be seen that the modes have been drawn in all cases to indicate that the centre of gravity remains fixed and that no rotation occurs, as these motions have already been fully described by the other five degrees of freedom. v_1 and v_3 mainly involve C—H stretching, and hence are of similar frequency to each other. They have about twice the frequency of v_2 because the hydrogen atom is so much lighter than the carbon atom. Similarly, they are much more affected in frequency than is v_2 upon the substitution of deuterium for hydrogen. v_1 is of slightly higher frequency than v_3 because it involves the simultaneous change of all three bond lengths. The bending modes v_4 and v_5 are of considerably lower frequency, as would intuitively be expected, and are also of very similar frequency.

An example of the sort of deductions that are possible from a simple infrared spectrum is given in figure 9.5.

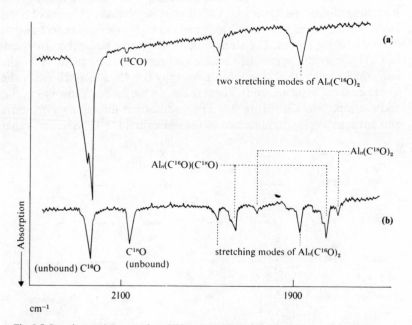

Fig. **9.5** Superimposed spectra of (a) $C^{16}O$ and (b) $C^{16}O$ with 40 °₀ $C^{18}O$ condensed in Kr matrix along with Al atoms. The spectra strongly support the presence of a non-linear dicarbonyl aluminium complex. (A. J. Hinchcliffe, J. S. Ogden, and D. D. Oswald, *Chem. Comm.*, 1972, 338.)

The normal modes of larger molecules (especially non-planar ones) are often hard or impossible to guess intuitively. However, there exists an elegant method of analysis using symmetry properties and group theory, called normal coordinate analysis, which is capable of predicting the normal modes of molecules of up to about twelve atoms, provided that the molecules have appreciable symmetry.

Still larger molecules usually have the simplifying advantage that different parts of the molecule may be considered to vibrate almost independently of most other parts. For example, an organic cyano group always shows a characteristic frequency known as the $-C\equiv N$ stretch, which varies only within a range of about 5%, whatever the molecule to which the cyano group is attached (even when it is conjugated). Tables of such 'group frequencies' are available,[4] which often make it possible to identify medium-sized molecules directly from their infrared spectrum. It should always be borne in mind, however, that group frequency descriptions such as 'C—C stretch' and 'C—H bend' are only approximate descriptions and do not exclude the possibility of some motion of other parts of the molecule being involved in the absorption.

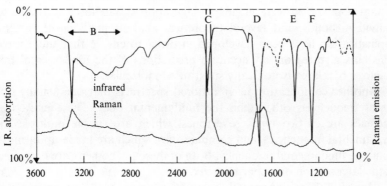

Fig. **9.6** Infrared and Raman spectra of propynoic acid HC≡C—CO.OH from 1000—3600 cm⁻¹ [5]

An even more direct analytical use of infrared (and, increasingly, Raman) spectra is to compare the spectrum of the unknown substance with one of the extensive commercially available compilations of spectra. This use of spectra is known as 'fingerprinting', and attempts are currently being made to cut down the labour of comparing the spectra of known and unknown substances by the use of computers. A typical infrared absorption spectrum is shown in figure 9.6.

[4] For example, A. C. Cross, *An Introduction to Practical Infrared Spectroscopy*, Butterworths, London, 1964. L. J. Bellamy, *The Infrared Spectra of Complex Molecules*, Methuen, London, 1958.
[5] J. E. Katon and N. T. McDevitt, *Spectrochim. Acta*, **21**, 1717, 1965.

The assignments of some strong absorptions, made with the help of deuterium substitution, are:

A C—H stretching mode
B O—H stretching mode, much broadened by H-bonding
C C≡C stretching mode (i.r. active because of molecular asymmetry)
D C=O stretching mode
E C—O stretching mode
F C—O—H bending mode (less affected by H-bonding)

The corresponding Raman spectrum is shown below the infrared spectrum. Absorptions A, C, D, E (very weak) and F may be identified. The exceptional strength of C may be attributable to the relatively high polarizability of the C≡C bond.

9.7 Theoretical analysis of vibrational spectra

The vibrations of polyatomic molecules cannot be analysed in terms of force constants in the simple way that is possible for diatomic molecules. This is because the stretching of one bond (especially in a conjugated molecule) appreciably affects the bonds nearest to it, and may affect more distant bonds. However, one may assign approximate force constants to individual bonds, and several theoretical analysis methods of varying sophistication have been developed in order to ensure that these force constants give reasonably accurate predictions of the fundamental frequencies of related isotopically substituted molecules.

Another complication in vibrational spectra is that one usually observes frequencies other than the fundamental ones. These extra frequencies are of two types: overtones, which are simple multiples of fundamental frequencies, and combinations, which are made up of more than one fundamental frequency. Both of these phenomena are discussed in the later section on anharmonicity.

However, there is one greatly simplifying fact in the theoretical analysis of infrared and Raman spectra of molecules. This is that the only vibrations whose fundamental $0 \rightarrow 1$ absorptions give rise to infrared absorption lines are those which involve a fluctuating electric dipole. This rule has already been discussed for diatomic molecules. For polyatomics it is immaterial whether the dipole fluctuates around a constant value or around zero. For example, the v_2 and v_3 modes of CO_2, and the v_3 and v_5 modes of C_2H_2, are infrared active, because the bent or distorted molecule has a dipole moment, whereas the other modes are not active, and do not appear in the infrared spectrum. This rule is adhered to strictly for isolated molecules, but breaks down upon isotopic substitution (e.g. H—C—C—D, when v_1, v_2, and v_4 also appear in the infrared spectrum) and even in the gas at fairly high pressure, where temporary association of molecules is significant.

The infrared gross selection rule is only of relevance to centrosymmetric and a few other highly symmetric molecules as all the vibrations of other molecules are virtually bound to involve fluctuating dipole moments. It is a particularly powerful rule when it applies, because its converse also applies strictly to Raman spectra. In other words, the infrared active fundamentals of a centrosymmetric molecule never appear in the Raman spectrum, and vice versa. This is known as the mutual exclusion rule, and is discussed further in appendix 1. An approximate reason for the existence of Raman inactive modes is given by the classical theory of Raman absorption (section 9.4), from which one sees that only vibrations which change the molecular polarizability (i.e. roughly, the volume of the molecule) can give rise to Raman lines.

An additional experimental aid in the analysis of vibrational spectra is the use of polarized radiation. In infrared absorption, the radiation must clearly be polarized in the direction of the fluctuating dipole in order for absorption to take place. If the absorbing molecules are oriented, e.g. in a crystal, the various absorptions may be distinguished by the various angles that the fluctuating dipole makes with the main molecular axis.

In Raman measurements it is usual for the exciting radiation to be plane polarized, especially if it is laser radiation. The Raman scattered radiation will usually remain largely polarized, as predicted in chapter 3 for Rayleigh scattering, if the scattering vibration is a breathing mode, but largely unpolarized (or depolarized) if it is any other mode. There is a theoretical limit which means that no vibration can ever depolarize the scattered radiation by a factor of more than $6/7$.[6]

9.8   Effects of anharmonicity

General

So far it has been assumed that molecular vibrations are simple harmonic oscillators, and that the potential well in which a nucleus is held is parabolic in any given direction. This is a fairly substantial approximation, as may be seen from figure 9.1. If the potential well has any shape other than that of a parabola then the separation ΔE between the various vibrational energy levels will no longer be simply $h\upsilon_{\mathrm{vib}}$. For example, the vibrational levels for the real potential in figure 9.1 gradually converge in energy, so that there is an infinite number of them below the dissociation level. The exact calculation of these vibrational energies depends on an exact knowledge of the potential well. An approximation which usually works quite well for the lower levels is

$$\Delta E = (n+\tfrac{1}{2})h\upsilon_{\mathrm{vib}} - (n+\tfrac{1}{2})^2 hx\upsilon_{\mathrm{vib}} \tag{9.9}$$

This is actually an exact solution of the wave equation for the 'Morse

[6] See, for example, T. R. Gilson and P. J. Hendra, *Laser Raman Spectroscopy*, John Wiley (Interscience), New York, 1970.

potential', $PE = D[1 - e^{-a(R-R_e)}]^2$, which is quite a good approximation to a real potential. x is related to D and a and is a measure of the anharmonicity. The physical meaning of v_{vib} with an anharmonic potential is the classical frequency of vibration as one approaches zero vibrational energy (a situation which is never actually realized because of zero point energy). The actual frequency of the molecular vibration will depend on the vibrational amplitude.

When anharmonicity is included, the strict selection rule $\Delta n = \pm 1$ becomes only a partial selection rule, and other transitions, known as overtones, become weakly allowed. Furthermore, the $\Delta n = \pm 1$ transition splits into several lines ('hot bands') at high temperatures as the population of the higher vibrational levels becomes significant. This is because transitions from levels $1 \rightarrow 2$, $2 \rightarrow 3$, etc. of an anharmonic oscillator differ slightly in energy from the usual $0 \rightarrow 1$ transition.

It is sometimes possible to work backwards from a knowledge of the vibrational energy levels to deduce the shape of the vibrational potential well. This is one way of distinguishing symmetrical from unsymmetrical hydrogen bonds. In a symmetrical potential well the strict selection rule $\Delta n = \pm$ odd number holds (see problem 9.8) despite anharmonicity.

Polyatomic molecules

Anharmonicity has a further effect on polyatomic molecules. We saw in problem 2.5 that molecular vibrational motions could only be analysed in terms of normal modes provided that the motion of each bond in the molecule obeyed Hooke's law. For real molecules this is not true, and at first sight it would appear that the useful concept of normal modes has to be abandoned. However, this is fortunately not really necessary, if we redefine the normal modes of the real molecule as the (truly) normal modes in the (unattainable) limit of zero vibrational amplitude. We are then able to analyse any vibration with finite amplitude into contributions from these truly normal modes; the only difference from the harmonic case is that no real, i.e. anharmonic, vibration is possible which is in itself a pure normal mode, rather than a combination. In quantum mechanical terms, one finds that the transformation from Cartesian to normal co-ordinates no longer allows full separation of the wave equation into one-dimensional equations. The physical consequence of this is that new 'combination' bands appear in the vibration spectrum, corresponding approximately to simultaneous absorption of radiation by more than one normal mode. Fortunately, these combination absorptions are usually weaker than the fundamental absorptions.

It is not immediately obvious in what way the anharmonic vibrations can be partially like normal modes and partially unlike them. The example of the v_3 vibration of carbon dioxide, redrawn in figure 9.7, may make matters slightly clearer. In (b), each CO bond is increasingly hard to compress, and increasingly easy to stretch in comparison with a bond obeying Hooke's law. This is equivalent to superimposing a motion (c) onto the normal harmonic motion (a). (c) is not a harmonic motion, for it only has significant amplitude at large amplitudes of (b). It is in fact the extra anharmonic part of the symmetrical v_1 vibration. Thus the v_2 and v_1 modes are linked by a common extra anharmonic part which can be shown to make simultaneous v_1 and v_2 transitions fairly likely.

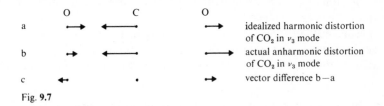

Fig. **9.7**

Fermi resonance

The overtone and combination bands described above are sometimes found to be shifted in frequency due to an interaction known as Fermi resonance. For example, the v_1 and v_2 modes of carbon dioxide occur at 1337 and 667 cm^{-1} respectively. One would expect therefore that the first v_2 overtone would appear at 1334 cm^{-1}, very close to the v_1 fundamental. In fact, neither of these lines appears in the Raman spectrum, but instead one observes quite strong Raman lines at 1388.3 and 1285.5 cm^{-1} due to modes which are combinations of v_1 and v_2.

The reason for this unexpected complication is that our analysis so far, involving separation of normal coordinates, is not valid under the special conditions of vibrational coordinates which are not precisely normal (due to anharmonicity) and which have frequencies which are approximately multiples of each other. The quantum mechanical argument for Fermi resonance is somewhat analogous to the one which explains predissociation (problem 10.11) and also the separation of the $I = 1, m_I = 0$ nuclear spin state from the $I = 0, m_I = 0$ state in molecular hydrogen under the influence of hyperfine coupling. It is quite common for a minor effect (such as anharmonicity or hyperfine coupling) to have a much enhanced influence on certain specific energy levels when the major effects (such as force constants or Zeeman interaction) which normally determine these levels lead by themselves to an accidental near-degeneracy of these specific levels. This influence is nearly always such as to widen the energy gap between the near-degenerate levels, often symmetrically.

Classical mechanics provides a partial insight into Fermi resonance. Consider the example of carbon dioxide quoted above. At high amplitudes the v_2 (bending) mode can be assisted by a simultaneous v_1 stretch at twice the frequency which has the effect of weakening the bond at maximum stretch and hence of weakening the molecular rigidity at the extremes of the bending mode. Thus an in-phase combination of v_1 with v_2 will greatly heighten the effect of anharmonicity on v_2 and thus lower its frequency at high vibrational amplitudes. This will make the first overtone of v_2 much stronger than it would otherwise have been (because of its association with a fundamental) and of lower frequency (1285.5 instead of $1334\,cm^{-1}$). Symmetry requires that an out-of-phase combination, of increased frequency and similarly enhanced intensity, should also exist. This explains the second line at $1388.3\,cm^{-1}$. It will be noted that the vibrations v_1, v_2 have compatible symmetry; this is a normal requirement for Fermi resonance to occur.

Fermi resonance occurs quite commonly in the vibration spectra of larger molecules, and because of this, one is unfortunately not always entitled to deduce that two strong lines imply two distinct fundamentals.

9.9   Simultaneous rotational and vibrational transitions

The discussion of absorption of radiation energy by vibrations has so far made the simplification that the radiation was not quantized. However, at the end of chapter 4 it was pointed out that radiation is quantized, and that the photons carry an angular momentum of $\sqrt{2}\hbar$, with allowed z-components of $\pm\hbar$ (but not of zero). When a photon is absorbed by a body, this angular momentum is transferred to the body. The resulting angular momentum (i.e. rotational state) may be calculated either by analogy with the problem of adding electron orbital angular momenta (cf. chapter 10), or by analysis of the transition moment, and in either case the conclusion is that the rotational quantum number J and the orientational quantum number m_J simultaneously change by ± 1. In the case of the vibrational spectra of condensed phases this has no observable effect, because the angular momentum is almost immediately transferred to or from the lattice surrounding the absorbing molecule. However, in low pressure gas phase vibration spectroscopy it has considerable effect, as may be seen from the high resolution infrared spectrum of HCl gas in figure 9.8. The left-hand portion of the spectrum is known as the

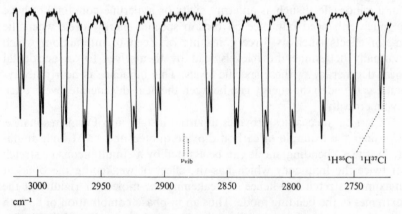

Fig. **9.8** High resolution infrared absorption spectrum of HCl gas (part)

R-branch, and shows fully absorptive transitions governed by the selection rule $\Delta n = +1$, $\Delta J = +1$. It is almost exactly the same as the pure rotation spectrum of HCl, shifted up in frequency by the vibrational frequency v_{vib}. The right-hand portion is known as the P-branch. It is governed by $\Delta n = +1$, $\Delta J = -1$ and is thus the rotational emission spectrum of HCl shifted up in frequency v_{vib}. The line separation increases in the P-branch and decreases in the R-branch for exactly the same reason, namely centrifugal distortion, as it does in the pure rotational spectrum.

However, there is a second reason why the lines are not evenly spaced which is often of more importance than centrifugal distortion. This is that the moment of inertia of a molecule (which determines the line spacings) will usually be slightly different in different vibrational states. When vibration is considered the formula $I = \Sigma mr^2$ (8.4) for the moment of inertia must be understood to imply summation over all possible nuclear positions, suitably weighted, i.e. over the spread of the nuclear vibrational wavefunctions. The extra factor r^2 means that the far-flung temporary positions of the nuclei contribute disproportionately greatly to the moment of inertia. For a diatomic molecule these far-flung positions are increasingly likely as n increases, and therefore the moment of inertia increases with n. For polyatomic molecules it is also capable, in some circumstances, of decreasing with increasing n. If the moment of inertia in the state $n = 0$ is I_0 and in the state $n = 1$ is I_1, then the change in rotational energy for the transition from level J to $J+1$ (for example) will therefore be

$$\frac{\hbar^2}{2}\left[\frac{(J+2)(J+1)}{I_1} - \frac{J(J+1)}{I_0}\right] \tag{9.10}$$

which will differ slightly from the value $\hbar^2(J+1)/I$ (equation 8.2) predicted for a pure rotational transition, and will enhance the convergence or divergence of the P- and R-branches. If the spectrum can be fully analysed, it is possible to extract both I_0 and I_1, and to compare these with the values predicted from the known vibrational wavefunctions in order to refine the data.

Diatomic species with non-zero electronic angular momentum, such as nitric oxide, may also undergo vibrational transitions where $\Delta J = 0$. In these cases the quantum of angular momentum affects the electronic rather than the nuclear motion. The $\Delta J = 0$ transitions give rise to a very much more compact group of lines, based on the fundamental frequency v_{vib} and called a Q-branch. Indeed, if $I_n = I_0$ the Q-branch is simply a single line at v_{vib}.

The Raman spectra of gases, where they have been observed, show similar rotational fine structure. Because the Raman effect is a two-quantum process, it follows that the selection rules are $\Delta J = \pm 2$ or 0, and therefore that a Q-branch will always occur. The $\Delta J = +2$ branch is known as an S-branch and the $\Delta J = -2$ as an O-branch.

In polyatomic molecules very similar effects are observed. However, there are additional complications, partly because there is a wide variety of ways in which molecules can accept or lose a quantum of angular momentum. A non-linear molecule, for example, may change any of its three components of angular momentum, and hence the two components described by the quantum number J may be unaffected, giving rise to a Q-branch. Even a linear molecule may have a Q-branch, because any two degenerate bending vibrations may combine exactly 90° out of phase so as to give the molecule an effective rotation about its axis (see figure 9.9). Further complications arise when electron and nuclear spin are taken into account. The presence or absence of a Q-branch is information of value in enabling a spectroscopist to determine the type of vibration he has recorded, particularly as the Q-branch may often be identified under only moderate resolution. Vibration-rotation spectra provide a most valuable alternative to pure rotation microwave spectra for obtaining molecular moments of inertia, as they are often easier to observe. Indeed, in some cases (e.g. CO_2) the pure rotational spectrum is unobtainable because the molecule lacks a dipole moment.[7]

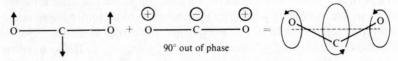

Fig. **9.9** Example of axial rotation of a linear molecule made possible by bending distortion

9.10 Linewidths

The linewidths of bands in both infrared and Raman spectra are usually determined by the extent of collapse of the rotational fine structure due to intermolecular collisions (as described under 'pressure broadening' in section 6.2(iv)). In liquids the rotational structure is usually not quite fully averaged away. Further broadening can occur due to strong inter-molecular interactions. When these occur one must consider the vibra-tions of the whole lattice, as in the Debye treatment of the specific heats of solids, rather than those of isolated molecules. A notorious example of a very broad line is the bending vibration of H_2O in wet samples, which is strong enough to obscure most of the infrared region below about $800 \, cm^{-1}$, and which contributes to the heat-blanketing effect of terrestrial cloud layers.

9.11 Additional effects in Raman spectroscopy

The theory of Raman spectroscopy given so far has been confined to moderate intensities of incident radiation, at frequencies well removed

[7] The classic text covering such matters in detail is: G. Herzberg, *Infrared and Raman Spectra*, Van Nostrand/Reinhold, Princeton, 1945.

from the optical absorption frequencies of the sample concerned. The advent of high power laser Raman spectroscopy has made it possible to observe the effects of high incident radiation intensity, and to obtain acceptable Raman spectra from light-absorbing samples. The special effects observed include:

a *Resonance Raman scattering*
The classical scattering theory of chapter 3 predicted very great scattering of radiation near an absorption frequency of a sample. The Raman scattering of the radiation is similarly greatly enhanced in these circumstances, and the relative line intensities are also affected.

b *Hyper-Raman scattering*
When very high radiation intensities are used, then equation 3.1, which defines the isotropic polarizability, α, has to be replaced by the equation

$$\mu = \alpha E + \beta E^2 \tag{9.11}$$

β, which is usually negative, is known as the hyperpolarizability, and its presence allows for the fact that there is a limit to the extent to which linear distortion of electron orbits by an electric field is possible. If the classical treatment of section 9.4 is applied using equation 9.11 instead of the first part of equation 9.4, then a component of scattering at $2v_0$ is predicted proportional to β, together with terms at $2v_0 \pm v_{\text{vib}}$, proportional to $d\beta/dR$. The scattering at $2v_0$ is known as frequency doubling, and this phenomenon is used to extend the available frequencies of laser radiation. The 'hyper-Raman' scattering at $2v_0 \pm v_{\text{vib}}$ offers a third type of vibration spectroscopy with its own selection rules, etc., which is able to supplement the information gained from normal Raman and infrared spectra. However, it is still very much in a stage of development.

Suggestions for further reading

A. D. Cross, *An Introduction to Practical Infra-red Spectroscopy*, Butterworths, London, 1964.

M. Davies (ed.), *Infra-Red Spectroscopy and Molecular Structure*, Elsevier, London, 1963.

G. Herzberg, *Molecular Spectra and Molecular Structure*, Van Nostrand/Reinhold, Princeton, N.J., 1945, vols I and II.

T. R. Gilson and P. J. Hendra, *Laser Raman Spectroscopy*, John Wiley (Interscience), New York, 1970.

L. J. Bellamy, *The Infra-Red Spectra of Complex Molecules*, Methuen, London, 1958.

A. Finch, P. N. Gates, K. Radcliffe, F. N. Dickinson, and F. F. Bentley, *Chemical Applications of Far Infrared Spectroscopy*, Academic Press, London, 1970.

Problems

9.1 Plot V against x for the Morse potential $V = D(1 - e^{-a(x - x_e)})^2$ when $D = 200 \text{ kJ mol}^{-1}$, $x_e = 0.1 \text{ nm}$ and $a = 20 \text{ nm}^{-1}$.

Show that a particle whose energy is near the minimum of this potential well obeys Hooke's law fairly closely.

9.2   A particle of mass m at position x has a potential energy $V = K(x - x_e)^2$, where K and x_e are constants.

   a Show that, under the laws of classical mechanics, $K = 2\pi^2 m v_{vib}$, where v_{vib} is the particle's vibration frequency.

   b Show that if the particle has a total energy of $\frac{1}{2}h v_{vib}$, and obeys the laws of classical mechanics, then its total range of motion (which is roughly twice the uncertainty in position Δx) depends inversely upon the square root of v_{vib} and of m.

   c In quantum mechanics the particle may be considered to have an uncertainty in velocity approximately equal to its maximum velocity (at $V = 0$). Using the uncertainty principle, show that, on this analysis also, Δx is inversely proportional to the square root of v_{vib} and of m.

9.3   **a** Using table 9.1, calculate the enthalpy change for the reaction $H_2 + D_2 = 2HD$ assuming that only ground vibrational states are occupied. What would be the effect of also taking molecular rotation into consideration?

   b Justify the zero-point energy correction for HD, using the reduced mass expression 9.8.

9.4   Nitrous oxide gas, N_2O, has an infrared spectrum with absorptions centred on $588.8 \, cm^{-1}$, $1285.0 \, cm^{-1}$ (no Q-branch) and $2223.5 \, cm^{-1}$ (no Q-branch). The latter two frequencies have also been clearly identified in the Raman spectrum. What can you deduce from this about the structure of N_2O?

9.5   **a** NO_2 has an ONO bond angle of $134°$. Sketch its three normal modes of vibration as accurately as you can. (They have frequencies of 648, 1320, and $1621 \, cm^{-1}$ and it may help to consider them as being derived from the normal modes of CO_2.)

   b N_2O_4 may be considered as two NO_2 molecules weakly linked by the nitrogen atoms, viz.:

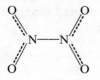

Using the known vibrational and rotational modes of NO_2 and considering their in and out-of-phase combinations, deduce (approximately) the twelve vibrational modes of N_2O_4. Which of these would you expect to be infrared active?

9.6   When a certain molecule is placed in an electric field E it acquires an induced dipole moment μ given by the formula

$$\mu = \alpha E + \beta E^2$$

α is the polarizability, and depends on time according to the equation

$$\alpha = \alpha_0 + a \cos 2\pi v_t$$

where a is a constant and v a vibration frequency. Similarly β is the hyperpolarizability, and varies with r so that $\beta = \beta_0 + b \cos 2\pi vt$. Prove that if $E = E_0 \cos 2\pi v_0 t$, the frequencies v_0, $v_0 \pm v$, $2v_0$ and $2v_0 \pm v$ will all appear in the spectrum of the scattered light.

9.7   The infrared absorption spectrum of $^1H^{35}Cl$ gas has no line at the vibration
frequency $2885.90\,cm^{-1}$ (corresponding to the $n = 0 \rightarrow 1$ transition). The
nearest lines on either side are at 2865.09 and at $2906.05\,cm^{-1}$. Explain this,
and calculate the percentage change in bond length between the $n = 0$ and
$n = 1$ vibrational states, ignoring anharmonicity and centrifugal distortion.

9.8   **a** Using appendix 1, prove the selection rule $\Delta n = \pm$(an odd number), for a
symmetrical but non-parabolic potential well.

   b It has been suggested that the blue colour of clear water may be due to
absorption by an O—H stretching vibration overtone. Discuss this. The
blue colour of thick ice is slightly different from that of water. Suggest a
reason why this might be so.

10 Visible and ultraviolet spectroscopy

10.1 Introduction

There are four distinguishable types of visible and ultraviolet spectroscopy, which may be roughly described as the absorption and emission spectroscopy of gaseous atoms, the absorption spectroscopy of gaseous molecules, that of liquids and solutions, and that of solids. Unexpectedly, the spectra of liquids and solutions are in many ways the simplest to understand, and those of atoms the hardest.

The spectra of solutions have two simplifying features. The first is that many of the static intra- and inter-molecular interactions that complicate gaseous and solid state spectra become dynamic interactions in the liquid state. Instead of contributing to line splittings they merely cause line broadening and small frequency shifts. Therefore it is usually possible to identify at least the long wavelength absorptions in the spectra of liquids as due to transitions to excited states that are predicted in principle (if not in precise energy) by elementary bonding theory. The second simplifying feature is that, because the liquids are usually at or near room temperature, observed transitions are almost invariably upwards from the ground state only. The order of absorption peaks therefore corresponds to the order of excited states to which transitions from the ground state are permitted.

This second feature also simplifies solid state spectra. But many of the intermolecular interactions in solids are static, and therefore these must be included in the energy level calculations in order to explain the observed absorptions, which are generally rather sharper than in liquids, and of greater complexity. The vapour phase spectra of molecules are even more complex, because the molecules stay for long enough in particular vibrational and even rotational states to permit the identification of these from the spectra. Compared with those of liquids, the u.v. and visible spectra of gases are usually very highly resolvable, and consist of series of bands each corresponding to one electronic transition but a wide variety of upper and lower vibrational and rotational states. Such spectra are often difficult to analyse, but when full analysis is possible it can yield all the information offered by rotational and vibrational absorption spectroscopy, not only for the ground state but also for some excited states, together with dissociation energies. This sort of detailed information on bond lengths, angles, and strengths has greatly extended the inorganic and theoretical chemist's understanding of the bonding in small molecules. Figure 10.9 shows a comparison of the vapour, liquid, and crystalline phase spectra of bromobenzene.

The spectra of very small molecules and of atoms do not show such complex vibrational and rotational fine structure. However they usually reveal an even greater complexity of electronic states. The reasons for this will be explained later; briefly, the complexity arises from the fact that electrons in many small molecules and (even more so) in atoms commonly exist in states whose orbital and spin angular momentum are non-zero. The interactions between electrons in such states are much more complex than in the states with zero average electron angular momentum which are commonly found in large molecules, and therefore there are far more possible energy levels to consider when analysing the absorption or emission spectrum.

Fortunately, although the theory of the electron energy levels involved in visible and ultraviolet spectroscopy is hard to understand, the experiments are usually easy to perform. As is explained in chapters 5 and 6, high signal to noise ratio and resolving power are relatively easy to obtain, and the main experimental limitation is often that of obtaining a sufficiently small linewidth from a given sample. For this reason visible and ultraviolet spectroscopy is a powerful tool in the observation and study of low concentration and transient species, as in flash photolysis. The simplicity of the technique also permits its use by astronomers in the study of stellar, interstellar, and planetary gases (figure 5.2).

10.2 Perturbation theory

We must now consider the notoriously complex problem of predicting atomic[1] and molecular electron states. The energy of an atom or molecule with n nuclei plus electrons depends on the interplay of $n(n-1)$ interactions, each of which will be both Coulombic and magnetic. The two-particle problem of the hydrogen atom was partly solved in chapter 2. The three-particle problem of the helium atom is about an order of magnitude more difficult, but has been substantially solved. Each additional particle beyond this number adds a further order of magnitude to the difficulty of the problem, and it has been estimated that even if the exact wavefunction for a heavy atom such as Cs could be calculated, the observable universe would probably not be large enough to store that information. It follows that the calculation of electron energy levels in atoms heavier than helium involves much judicious approximation.

One of the most useful approximation techniques for solving complex equations in the physical sciences is perturbation theory. This theory can be stated in several forms,[2] appropriate to different problems, and was

---

[1] Those wishing to know only the results of this problem and not the reasoning involved should omit the next four sections and consult an elementary text under the heading 'term symbols'.

[2] A concise statement of a form of perturbation theory well suited to systems with a limited number of quantum states, as in magnetic resonance, is given in R. N. Dixon, *Spectroscopy and Structure*, Methuen, London, 1965, p. 19.

first developed in the eighteenth century by astronomers who wished to calculate the small deflections from elliptical planetary orbits caused by inter-planetary gravitational attractions. The form in which it will be developed in the next few sections is particularly appropriate to the problem of predicting atomic electron states.

The essence of perturbation theory is to simplify the equations involved in the physical problem under consideration until they are completely solvable, and then to treat the interactions that are ignored in this simplification as 'perturbations', or small correction terms, which only have a relatively small effect on the solution of the problem. We will see that it is often necessary to choose the method of approximation quite carefully in order that the perturbations do have only a small effect on the solutions.

In quantum mechanical problems one normally has to attempt a solution of a Schrödinger equation $(H_0 + P)\psi = E\psi$, where one already knows the solutions ψ_k of the simpler equation $H_0\psi = E\psi$, and the corresponding eigenvalues E_k. The problem is to find the solutions ψ in the presence of the extra perturbation term P. The central assumption of perturbation theory is that if $P \ll H_0$, the various perturbed states ψ do not differ greatly from the unperturbed states ψ_k. Thus one imagines the perturbation P being gradually increased from zero to its full value, causing each state ψ_k to turn gradually into a rather similar state $\psi(k)$.

Using this assumption, and also the general fact that any particular state $\psi(n)$ can always be written as a series in terms of the complete set of states ψ_k, it is fairly easy (appendix 2) to derive the following two equations giving the energy and the wavefunctions, $\psi(n)$, of a particular perturbed state derived from the unperturbed state ψ_n, in terms of P, E_n, E_k, and ψ_k:[3]

$$\text{Energy difference between } \psi(n) \text{ and } \psi_n = \int \psi_n^* P \psi_n \, d\tau \tag{10.1}$$

$$\psi(n) = \psi_n - \sum_{k \neq n} \int \frac{\psi_k^* P \psi_n \, d\tau}{E_k - E_n} \psi_k \tag{10.2}$$

In equation 10.2, E_k and E_n are the eigenvalues of ψ_k and ψ_n with respect to H_0, and the ψ_k are assumed to be normalized. The solution $\psi(n)$ is almost as good as an exact solution of the wave equation provided that our original assumptions are valid. Now equation 10.2 is consistent with the original assumption that $\psi(n) \approx \psi_n$, provided that $E_k \not\approx E_n$. In other words, a small extra potential energy term P in the Schrödinger equation does not affect the solutions greatly as long as they are completely non-degenerate; and only affects their energies moderately, as given by equation 10.1. So perturbation theory predicts reliable approximate wavefunctions provided that the initial unperturbed states are not even ap-

---

[3] It is important to distinguish k, which is a general label for all the states involved, from n, which is the label of any one specific state.

proximately degenerate. However, normally some of the unperturbed states will be degenerate, and so we also need to discover the conditions under which perturbation theory remains valid for degenerate states.

Let us consider what would happen if all the ψ_k were completely non-degenerate except for two states, ψ_n and $\psi_{n'}$. Equations 10.1 and 10.2 would apply to all the ψ_k except for these. But we apparently know nothing about the effect of the perturbation on ψ_n and $\psi_{n'}$ because in this case our perturbation theory as it stands is not self-consistent. However, we could argue as follows. Under perturbation, ψ_n and $\psi_{n'}$ must change into new states which are expressible as some combination of the ψ_k. Furthermore, because the perturbed $\psi(k)$ must all be orthogonal to each other, and because all the other $\psi(k)$ are almost the same as the ψ_k from which they derive, the only state substantially available for combination with ψ_n is $\psi_{n'}$, and vice versa. For if any other ψ_k were significantly represented in the perturbed states $\psi(n)$ and $\psi(n')$ then these could no longer be orthogonal to the state $\psi(k) \approx \psi_k$ ($k \neq n, n'$). Hence we have discovered that, under a weak perturbation, degenerate states combine with each other but not significantly with further states of substantially different energy. Let us call the correct combination states ψ_c, $\psi_{c'}$.

The next question is 'how do they combine?' This is not an easy question to answer for perturbations in general, although it is quite generally true that if combinations can be found which make all the integrals of type $\int \psi_c^* P \psi_c \, d\tau = 0$ then these are the correct ones. For when this is the case, ψ_c and $\psi_{c'}$ are not greatly affected (according to equation 10.2) by the perturbation and hence our perturbation theory is once again valid. Fortunately, a convenient method is available for finding correct 'diagonalizing' combinations[4] ψ_c, $\psi_{c'}$ in most atomic wavefunction calculations. The method is to find a set of mutually measurable atomic properties which are known to be unaffected by the perturbation. These properties and the perturbation must necessarily be mutually compatible, and therefore there must be some state of the real system which is simultaneously an eigenstate of all the operators corresponding to the measurable properties and also of the operator corresponding to the perturbation. There is no other way in which one could be free to measure any of the properties or to apply the perturbation at will, for only if $A_{(op)}\psi = \text{constant} \times \psi$, is ψ unaffected by all measurements or perturbations $A_{(op)}$. It follows that, if unique orthogonal combination wavefunctions, e.g. ψ_c and $\psi_{c'}$, can be found which are eigenfunctions of all the compatible measurements in the set, then these combinations will also be eigenfunctions of the perturbation, so that $\int \psi_c^* P \psi_{c'} \, d\tau$ will be zero.

Two such sets of compatible measurements are commonly considered for light atoms, namely $J_{(op)}^2$, $J_{z(op)}$, and $L_{(op)}^2$, $L_{z(op)}$, $S_{(op)}^2$, $S_{z(op)}$. The first

---

[4] The combinations ψ_c and $\psi_{c'}$ are said to 'diagonalize the perturbation' because if all the integrals such as $\int \psi_c^* P \psi_{c'} d\tau$ are written out in a square table or 'matrix', and if all the $\int \psi_c^* P \psi_{c'} d\tau = 0$ unless $c = c'$, the table will only have non-zero entries along its downward sloping diagonal. The diagonal entries will be the perturbation energies of equation 10.1.

of these sets is compatible with any perturbation which does not affect total angular momentum, and in particular with any purely internal interaction between the constituent particles of the atom, such as Coulombic repulsion. The second set is compatible with any perturbation which affects neither total spin nor total angular momentum. It is thus less generally valid than the first set, but more helpful where it does apply because it contains four operators and is thus more likely to produce a unique set of combinations ψ_c, $\psi_{c'}$ than is the first set with only two operators.[5]

A weak external magnetic field is an example of a perturbation that affects neither total nor orbital nor spin angular momentum (so long as each is already quantized along the field direction). If an otherwise isolated H atom is subjected to such a field in the z-direction, its orbital levels split under the perturbation after the manner shown in figure 10.1. (The further splittings due to spin are omitted at present.)

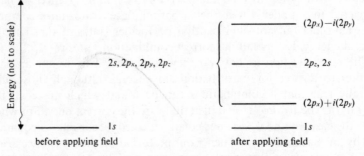

Fig. **10.1** Effect of weak magnetic field on lower levels of spinless H atom

The reason for this is as follows. Both before and after the application of the magnetic field the H atom is, to a very good approximation, in an eigenstate of $l^2_{(op)}$, $l_{z(op)}$, $s^2_{(op)}$, and $s_{z(op)}$ (note that $L = l$ and $S = s$ for a one-electron atom). Now all the eigenfunctions corresponding to 1s, 2s, 2p are eigenstates of $l^2_{(op)}$, but $2p_x$ and $2p_y$ are not eigenstates of $l_{z(op)} = -i\hbar(\partial/\partial\phi)$. However, the combinations $(2p_x) \pm i(2p_y)$ are; explicitly, they are $\psi[(2p_x) \pm i(2p_y)] = re^{-r/2a}\sin\theta \cdot e^{\pm i\phi}$, the eigenvalues being $\pm\hbar$. Hence they are the correct combination eigenfunctions. (The actual perturbation operator in this case is proportional to $-i\hbar B_0(\partial/\partial\phi)$, for an external field B_0, and the resulting splittings, calculated using equation 10.1, are proportional to mB_0.)

So far we have relied upon physical insight to find those atomic properties which remain fixed, and hence observable, despite the perturbation. A more formal way of discovering such observables is to use the fact that the operators corresponding to them must commute with the perturbation operator. The general reason for this was outlined in chapter 1. We saw there that if several properties of a system are simul-

[5] The two sets are not completely compatible with each other because the measurement of J_z removes the total independence of L_z and S_z by defining one unique axis for both. In other words $J_{z(op)}$ does not commute with $L_{z(op)}$ or with $S_{z(op)}$.

taneously and precisely observable, then the order of such observations must be irrelevant, and therefore the operators corresponding to measurement of the properties must commute. Applying a perturbation is formally equivalent to measuring a property, and hence if a pre-perturbation observable commutes with the perturbation it must remain observable even after the perturbation has been applied. Its eigenstates must therefore be valid eigenstates despite the perturbation, and thus give us the correct combinations ψ_c, $\psi_{c'}$ of the unperturbed states, i.e. those for which $\int \psi_c^* P \psi_{c'} \, d\tau = 0$. We will now apply this general perturbation argument to polyelectronic atoms, where the approach is necessary in order to cope with electron-electron repulsions and other internal interactions.

10.3   Starting states and the Pauli exclusion principle

Our first task is to find a usable set of unperturbed states for the polyelectronic atom. We will call these starting states. We have already discovered the precise one-electron states of the hydrogen atom, and have extended their use by observing that in the absence of all forces other than those between each electron and the nucleus, these states are equally valid for one electron in the presence of any number of other electrons, so that the total many-electron wavefunction is the product of the individual electron wavefunctions. As far as it goes this is a correct statement. But if we now ask which electron is in which state, then in general we discover an ambiguity. For consider a two-electron atom or ion, with one electron in a state with wavefunction ψ_a and the other in a state with wavefunction ψ_b. If we try to specify which state the first electron (1) is in and which one the second (2) is in, we find two possibilities, $\psi_a(1).\psi_b(2)$ and $\psi_a(2).\psi_b(1)$. These are both valid solutions of the wave equation and therefore any linear combination of them,

$$\psi_{comb} = A(\psi_a(1).\psi_b(2)) \pm B(\psi_a(2).\psi_b(1))$$

is also apparently a valid solution. However, we can immediately see that one restriction on such linear combinations is that $\psi_{comb}^* . \psi_{comb}$ must be unaffected by the exchange of electron (1) and electron (2) because the electrons are in fact indistinguishable. This still leaves several possibilities, however such as $A = B$, $A = -B$, $A = \pm iB$. However, Pauli's exclusion principle further restricts all two-electron wavefunctions to the one possibility of $A = -B$ so that the only possible wavefunction is $\psi_{anti} = A(\psi_a(1).\psi_b(2) - \psi_a(2).\psi_b(1))$. In its more general form, the Pauli principle requires that, upon exchange of any two electrons in a polyelectronic combination wavefunction, ψ_{comb}, the result must be $-\psi_{comb}$. The correct ψ_{comb}, which we shall hereafter refer to as ψ_{anti}, is then said to be 'antisymmetric'. With three electrons, for example, in states ψ_a, ψ_b, ψ_c, the only possible ψ_{anti} is

$$\begin{aligned}
\psi_{anti} = \; & \psi_a(1).\psi_b(2).\psi_c(3) - \psi_a(1).\psi_b(3).\psi_c(2) \\
& + \psi_a(2).\psi_b(3).\psi_c(1) - \psi_a(2).\psi_b(1).\psi_c(3) \\
& + \psi_a(3).\psi_b(1).\psi_c(2) - \psi_a(3).\psi_b(2).\psi_c(1)
\end{aligned}$$

which will be seen to change into $-\psi_{\text{comb}}$ following any of the three possible electron exchanges $1 \rightleftarrows 2$, $2 \rightleftarrows 3$, or $1 \rightleftarrows 3$. ψ_{anti} may be more neatly written as the 'Slater' determinant

$$\psi_{\text{anti}} = \begin{vmatrix} \psi_a(1) & \psi_b(1) & \psi_c(1) \\ \psi_a(2) & \psi_b(2) & \psi_c(2) \\ \psi_a(3) & \psi_b(3) & \psi_c(3) \end{vmatrix}$$

One immediate conclusion from all this is that if two of our one-electron states (including spin) are identical, $\psi_{\text{anti}} = 0$. This puts the Pauli exclusion principle into its more familiar form, which is that 'no two electrons may have the same quantum state (including spin)'.

We may illustrate the importance of antisymmetry by considering the filled 1s level of the helium atom. There are two possible one-electron wavefunctions, $e^{-r/a}\alpha$ and $e^{-r/a}\beta$, where α and β are the (separable) spin wavefunctions. The combination functions constructed from $e^{-r/a}\alpha(1) \times e^{-r/a}\alpha(2)$ and $e^{-r/a}\beta(1).e^{-r/a}\beta(2)$ are forbidden by the Pauli principle, and only $\psi_{\text{anti}} = e^{-r/a}\alpha(1).e^{-r/a}\beta(2) - e^{-r/a}\alpha(2).e^{-r/a}\beta(1)$ is allowed. This may be factorized, giving $\psi_{\text{anti}} = e^{-r/a}(1).e^{-r/a}(2).(\alpha(1)\beta(2) - \alpha(2)\beta(1))$, and in fact the spin wavefunction may be shown to be the $S = 0$, $m_s = 0$ eigenfunction of the operators $S^2_{(\text{op})}$ and $S_{z(\text{op})}$ (see footnote 6). Electron states with only this $S = 0$ possibility are called closed shells.

The first excited configuration of helium involves one electron with wavefunction $\psi = e^{-r/a}.\alpha$ or $e^{-r/a}.\beta$ and the other with wavefunction $\psi = (1 - r/2a)e^{-r/2a}.\alpha$ or $(1 - r/2a)e^{-r/2a}.\beta$. Four antisymmetric combinations are now possible, which are (factorizing where possible):

$$\psi_p = \left[e^{-r/a}(1).\left(1 - \frac{r}{2a}\right)e^{-r/2a}(2) - e^{-r/a}(2).\left(1 - \frac{r}{2a}\right)e^{-r/2a}(1) \right]$$
$$\times \left[\alpha(1).\alpha(2) \right]$$

$$\psi_q = \left[e^{-r/a}(1).\left(1 - \frac{r}{2a}\right)e^{-r/2a}(2) - e^{-r/a}(2).\left(1 - \frac{r}{2a}\right)e^{-r/2a}(1) \right]$$
$$\times \left[\beta(1).\beta(2) \right]$$

$$\psi_r = e^{-r/a}.\alpha(1).\left(1 - \frac{r}{2a}\right)e^{-r/2a}.\beta(2) - e^{-r/a}.\alpha(2).\left(1 - \frac{r}{2a}\right)e^{-r/2a}.\beta(1)$$

$$\psi_s = e^{-r/a}.\beta(1).\left(1 - \frac{r}{2a}\right)e^{-r/2a}.\alpha(2) - e^{-r/a}.\beta(2).\left(1 - \frac{r}{2a}\right)e^{-r/2a}.\alpha(1)$$

These are a complete correct set of degenerate starting states corresponding to $1s^1 2s^1$. Because they are all antisymmetric, their combinations must also be antisymmetric, and hence from now on we do not need to worry about the Pauli exclusion principle in taking further combinations.

10.4 Combination states

We have already seen that our task is to find combinations of our starting states which are eigenfunctions of $J^2_{(op)}$ and $J_{z(op)}$. In order to discover these combinations it is helpful to consider first only the strong perturbation due to electron-electron repulsion, for this leaves spin and orbital motion still independent of each other, so that our initial combinations must be eigenfunctions of $L^2_{(op)}$, $L_{z(op)}$, $S^2_{(op)}$, and $S_{z(op)}$. Any weaker perturbations will, following our main argument, only have a big effect upon those initial combinations which remain degenerate; they will cause further combination of these.

Spin angular momentum only

Many of the principles of combination of starting states may be brought out by considering the four starting states, ψ_p, ψ_q, ψ_r, and ψ_s above. In each of these states there is no orbital angular momentum, and hence any combination of them is bound to be an eigenstate of both $L^2_{(op)}$ and $L_{z(op)}$, with eigenvalues of zero. It remains to find out which combinations are eigenvalues of $S^2_{(op)}$ and $S_{z(op)}$. The latter operator is particularly simple. The z-component of total spin angular momentum equals the sum of the z-components of the separate spins that make up the total, because total angular momentum is conserved in all its components upon the interaction of initially separate particles. Hence, in our example, $S_{z(op)}\psi_p = \hbar\psi_p$ (i.e. ψ_p contains two electrons each originally with $S_z = \frac{1}{2}\hbar$), $S_{z(op)}\psi_q = -\hbar\psi_q$, and $S_z\psi_r = S_z\psi_s = 0$. Thus only ψ_r and ψ_s can combine, because any other combination of ψ_p, ψ_q, ψ_r, and ψ_s could not be an eigenfunction of S_z. In general, only starting states with the same m_s value can combine together. The correct eigenstates of $S^2_{(op)}$ and $S_{z(op)}$ are in fact ψ_p, ψ_q, $\psi_r + \psi_s$, and $\psi_r - \psi_s$ and their eigenvalues with respect to $S^2_{(op)}$ are $2\hbar^2$, $2\hbar^2$, $2\hbar^2$ (all $S = 1$) and 0 ($S = 0$) respectively, although it requires knowledge of the detailed form of $S^2_{(op)}$ to complete the proof of this.[6] The combinations $\psi_r \pm \psi_s$ happen to factorize, as follows, although this will not be the case in general:

$$\psi_r + \psi_s = \left[e^{-r/a}(1).\left(1 - \frac{r}{2a}\right)e^{-r/2a}(2) - e^{-r/a}(2).\left(1 - \frac{r}{2a}\right)e^{-r/2a}(1)\right]$$
$$\times \left[\alpha(1).\beta(2) + \alpha(2).\beta(1)\right]$$

$$\psi_r - \psi_s = \left[e^{-r/a}(1).\left(1 - \frac{r}{2a}\right)e^{-r/2a}(2) + e^{-r/a}(2).\left(1 - \frac{r}{2a}\right)e^{-r/2a}(1)\right]$$
$$\times \left[\alpha(1).\beta(2) - \alpha(2).\beta(1)\right]$$

It will be seen that ψ_p, $\psi_r + \psi_s$, and ψ_q correspond to the $M_s = 1$, 0, and -1 states of an angular momentum of $\sqrt{2}\hbar$, i.e. with $S = 1$, and that they each have the same space part. They are collectively called a triplet state, and, because of their common space part, are degenerate in the absence of a magnetic field. $\psi_r - \psi_s$ is called a singlet state ($S = 0$, $M_s = 0$)

[6] For example, W. Kauzmann, *Quantum Chemistry*, Academic Press, New York, 1957, p. 250, and also problem 10.12.

and has a different space part. The relative energies of the singlet and triplet states under a perturbation P may be found from equation 10.1. It may quite generally be shown that triplet states lie in energy below their related singlet states, the reason, briefly, being that the $+$ sign in the space part of $\psi_r - \psi_s$ or its equivalent means physically that the two electrons are forced to lie rather near to each other with consequent repulsion (problem 10.13). In general, the higher the multiplicity (defined as $2S+1$) of a state, the lower its energy.

Independent spin and orbital angular momentum

All of the above principles apply equally well to cases where an orbital angular momentum is present which does not interact with spin motions. Let us consider, for example, the ground configuration[7] of the carbon atom, $1s^2 2s^2 2p^2$. Now we may ignore the $1s^2$ and $2s^2$ as being closed shells[8] and concentrate on the two 2p electrons and their mutual interactions. If we write

$$\psi_+ = \psi_{2p(x)} + i\psi_{2p(y)}, \ \psi_0 = \psi_{2p(z)} \ \text{ and } \ \psi_- = \psi_{2p(x)} - i\psi_{2p(y)}$$

our starting states will be the fifteen possible antisymmetrized products of $\psi_+\alpha$, $\psi_+\beta$, $\psi_0\alpha$, $\psi_0\beta$, $\psi_-\alpha$, $\psi_-\beta$, (e.g. $\psi_+\alpha(1).\psi_0\beta(2) - \psi_+\alpha(2).\psi_0\beta(1)$, with $m_L = 1$ and $m_S = 0$). These will combine, using our previous argument, provided they have the same m_L and m_S values. The resulting combinations once again happen to factorize into space and spin parts, separately eigenstates of $L^2_{(op)}$ and $S^2_{(op)}$ as well as of $L_{z(op)}$ and $S_{z(op)}$, as follows:

$L = 2$
Term symbol ^1D
$$
\left.
\begin{array}{ll}
m_L = 2 & \psi_+(1).\psi_+(2) \\
\quad 1 & \psi_+(1).\psi_0(2) + \psi_+(2).\psi_0(1) \\
\quad 0 & 2\psi_0(1).\psi_0(2) + \psi_+(1).\psi_-(2) + \psi_+(2).\psi_-(1) \\
-1 & \psi_-(1).\psi_0(2) + \psi_-(2).\psi_0(1) \\
-2 & \psi_-(1).\psi_-(2)
\end{array}
\right\}
\begin{array}{l}
\text{all multiplied by} \\
{[\alpha(1).\beta(2) - \alpha(2).\beta(1)]}
\end{array}
$$

$L = 1$
Term symbol ^3P
$$
\left.
\begin{array}{ll}
m_L = 1 & \psi_+(1).\psi_0(2) - \psi_+(2).\psi_0(1) \\
\quad 0 & \psi_+(1).\psi_-(2) - \psi_+(2).\psi_-(1) \\
\quad 1 & \psi_-(1).\psi_0(2) - \psi_-(2).\psi_0(1)
\end{array}
\right\}
\begin{array}{l}
\text{all times} \\
\text{any from}
\end{array}
\left.
\begin{array}{l}
\alpha(1).\alpha(2) \\
{[\alpha(1).\beta(2) + \alpha(2).\beta(1)]} \\
\beta(1).\beta(2)
\end{array}
\right\}
$$

[7] Strictly this is a 'configuration' rather than a state, because as we will see, it contains within it several states with different energies, corresponding to different mutual interactions of the 2p electrons.

[8] Every 2p 2s starting state will contain $\psi_{1s}\alpha(1)\psi_{1s}\beta(2)\psi_{2s}\alpha(3)\psi_{2s}\beta(4)$, with its permutations, as a fixed factor. This changes the energy of the overall state by the same amount for each separate starting state.

$L = 0$

Term symbol ^1S

$m_L = 0 \; [\psi_+(1).\psi_-(2) + \psi_+(2).\psi_-(1) - \psi_0(1).\psi_0(2)] \; [\alpha(1).\beta(2) - \alpha(2).\beta(1)]$

In this example, because of the convenient factorization, it is easy to see that the $L = 2$ state is a singlet, the $L = 1$ state a triplet and the $L = 0$ state a singlet, and this information is given in code form in the 'term symbols' that are added. The use of D,P,S to describe L-values of 2,1,0 is analogous to the use of d,p,s to describe one-electron states in terms of their l-values.

If the two electrons had been from different quantum shells, as in the excited configuration $1s^2 2s^2 2p^1 3p^1$, six more non-zero antisymmetrized starting states would have been possible, (e.g. $\psi_{2+}\alpha(1) \times \psi_{3+}\alpha(2) - \psi_{2+}\alpha(2).\psi_{3+}\alpha(1)$) and a further 21 combination states would have been found, corresponding to three new levels ^3D, ^1P, and ^3S. The existence of the six states ^1D, ^3P, ^1P, ^3P, ^1S, ^3S is also predicted empirically by the simple 'vector addition rule',[9] which, briefly says that when various angular momenta with quantum numbers j_i, m_i ($j \geqslant m \geqslant -j$), the largest j_i being j_0, are added together in all possible allowed mutual orientations, the resulting total angular momenta correspond to all possible quantum numbers J, m_J, where

$$\sum_{(\text{all } i)} j_i \geqslant J \geqslant 0 \quad \text{or} \quad \left(j_0 - \sum_{(\text{all } i \text{ except } 0)} j_i \right)$$

whichever is the greater, and $J \geqslant m_J \geqslant -J$. The vector addition rule is incapable, in its simple form, of identifying states whose existence is forbidden by the Pauli exclusion principle.

10.5 Effect of internal magnetic interactions

In addition to the relatively strong electron-electron repulsion forces in light atoms, there are weaker magnetic forces. Every electron spin and every orbital angular momentum motion produces a magnetic field, and these separate fields all interact with each other. Fortunately, the spin-spin and orbital-orbital interactions by themselves present no real problem, for they are subject to exactly the same argument as are the electron-electron repulsions: the combined spin and combined orbital angular momentum states remain valid wavefunctions despite any interactions between the individual electron motions, provided that nothing links spin and orbital motion. The only effect of these two new interactions is to complicate the perturbation operator that one uses in equation 10.1 in order to predict the perturbation energy.

The magnetic interactions between the spin and the orbital angular momenta, however, add a complication. They are described by the collective term 'spin-orbit coupling', and have the effect of partially removing

[9] For example, W. Heitler, *Elementary Wave Mechanics*, Oxford University Press, London, 1956, p. 63 et seq., and W. Kauzmann, op. cit., p. 346.

the degeneracy of states, such as ^{3}P, where both spin and orbital angular momentum are non-zero. The reasoning whereby the perturbed spin + orbit combination states may be deduced is much as before. We know that, even in the presence of spin-orbit coupling, total angular momentum and its components are still conserved, so that the correct combination states will be eigenfunctions of $J^2_{(op)}$ and $J_{z(op)}$. They will also in this case be eigenfunctions of $L^2_{(op)}$ and $S^2_{(op)}$, because spin-orbit interaction can only change the direction of the spin and orbital angular momenta, and not their magnitude. Also, our $L_{z(op)}$ and $S_{z(op)}$ eigenfunctions of the previous section will be suitable starting states, provided that the spin-orbit coupling is appreciably weaker than the electron-electron repulsion which created these L and S states. They will combine with other states having the same value of $J_z (= L_z + S_z)$. In fact we have already seen how to combine two orbital angular momenta with $l = 1$ to give L-values of 2, 1, and 0, and exactly the same combinations serve in this case for combining the nine starting states (3 orbital × 3 spin) for $L = 1$ and $S = 1$, i.e. for ^{3}P, to give three states, with $J = 2$ (five-fold degenerate), $J = 1$ (three-fold degenerate), and $J = 0$ (non-degenerate). The J values provide an additional identifying label for the total states, viz. ^{3}P$_2$, ^{3}P$_1$, ^{3}P$_0$. The lowest level in the case of carbon is ^{3}P$_0$; in general, the application of equation 10.1 shows that for shells less than half full the states have increasing energy with increasing J, and for shells more than half full they have decreasing energy with increasing J. The final levels for $1s^2 2s^2 2p^2$ carbon are illustrated in figure 10.2(a), together with some of the observed lines in the spectrum of atomic carbon (figure 10.2b). Note that in $1s^2 2s^2 2p^1 np^1$ ($n \neq 2$) the ^{1}P, ^{3}S, and ^{3}D states are allowed.

So far we have considered only the case where spin-orbit coupling is much weaker than electron-electron repulsion. The opposite is the case for heavy atoms, and if we try to apply the perturbation argument of this and the previous section to such atoms we find a serious difficulty. For our argument assumes that only degenerate states mix, and this assumption is based on the last terms in equation 10.2, $\int \psi_k^* P \psi_n \, d\tau / (E_k - E_n)$, being negligible if $E_k \not\approx E_n$. However, if $E_k - E_n$ is merely the perturbation energy $\int \psi_n^* P' \psi_n \, d\tau$ of some smaller perturbation $P' < P$, it is obvious that $\int \psi_k^* P \psi_n \, d\tau / (E_k - E_n)$ will not in general be negligible, and may well be quite large, so that these nearby states will also have to be included in our combinations. The way to avoid this difficulty is to reverse the order in which we consider the perturbations—i.e. to combine l and s first under spin-orbit coupling to give individual total electron angular momenta of $\hbar \sqrt{j(j+1)}$ and then to combine these momenta to get the total atomic angular momentum under the influence of electron-electron repulsion. This analysis is known as 'j-j coupling'.

Further complexities of atomic energy levels can arise because of the quantization of nuclear spin orientation. However, these are only observable under very high resolution in atomic spectra, and their effect is calculable following the general principles described above.

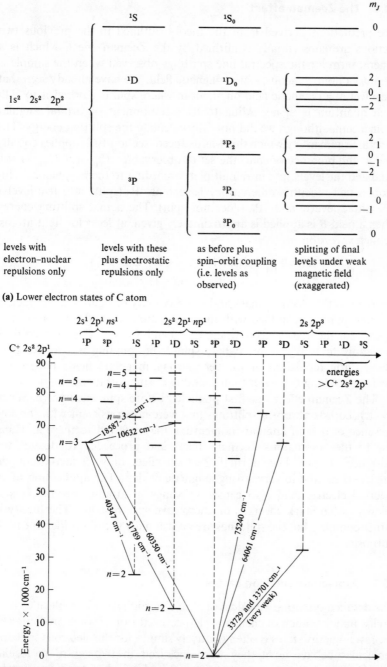

(a) Lower electron states of C atom

(b) Some observed absorptions of the C atom[10]

Fig. **10.2**

[10] A more complete 'term diagram' is available on p. 308 of H. C. Kuhn, *Atomic Spectra*, Longmans Green, London, 1962.

10.6 The Zeeman effect

The predictions derived from the theory outlined in the previous two sections are substantially confirmed by the Zeeman effect, which is a general term for the spectral line splittings observed when the sample is put into a homogeneous, static magnetic field. We have already seen that the effect of a magnetic field on an atom where spin is ignored is to split each quantum level according to its z-component of orbital angular momentum, although we did not fully calculate the splitting energy. The general argument by which this was deduced (section 10.2) applies equally well in the presence of spin: the set of observables $J^2_{(op)}$ and $J_{z(op)}$ is still valid and the levels split into multiplets according to their m_J values. This enables us to identify different levels such as $^3P_2$ (splits into five levels), $^3P_1$ (three levels), and $^3P_0$ (does not split). The actual splitting energy when a field B is applied is approximately given, at least for light atoms, by the expression:

$$\text{splitting} = \frac{ehm_J B}{2m_e c}\left(1 + \frac{J(J+1)+S(S+1)-L(L+1)}{2J(J+1)}\right) \tag{10.3}$$

where m_e is the electron mass and c the velocity of light. The number in brackets may be identified with the g of section 7.1, and may be seen to equal 2 for the free electron, where $L = 0$ and $S = \frac{1}{2}$.[11] The formula does not apply to heavy atoms (where spin-orbit coupling is stronger than electrostatic repulsion) or for very strong magnetic fields (where the Zeeman splitting is greater than that due to spin–orbit coupling).

The Zeeman effect was first predicted before spin was known. When spectroscopists observed splittings that were not consistent with the sole presence of orbital angular momentum, they referred to these as being due to the 'anomalous Zeeman effect', and proceeded to deduce the existence of spin. The 'normal' Zeeman effect only in fact occurs in singlet states, and in very strong magnetic fields.[12] The application of an external electric field also causes splittings in atomic spectra. This is known as the Stark effect, as in microwave spectroscopy. The theory is rather complex, but the splittings are occasionally useful for identification purposes.

10.7 Molecular electron states

The detailed quantitative calculation of electron states in molecules is in itself a major branch of chemistry. However, the qualitative prediction of the lower energy states is often relatively simple, for the electrons in many molecules behave as if they were localized in two-centre molecular orbitals, which are analogous to atomic orbitals but span two nuclei

---

[11] A partial derivation of 10.3 is given on·p. 269 of P. W. Atkins, *Molecular Quantum Mechanics*, Part III, Oxford University Press, London, 1970.

[12] ibid., p. 270. This 'decoupling' of spin is known as the Paschen–Back effect.

instead of being centred on just one. Which molecular orbitals will occur may be fairly reliably predicted from the principle of linear combination of atomic orbitals,[13] and these predictions can be extended to the more general case where the molecular orbitals are many-centred or delocalized. Many-electron molecular orbitals are derived like atomic orbitals, from considering the one-electron case and then making adjustments for electron-electron repulsions as further electrons are added according to the aufbau principle. However, the original molecular orbitals are not (usually) degenerate, and therefore first-order perturbation theory for non-degenerate states is (usually) valid; the one-electron m.o.'s merely need to be adjusted, rather than combined (except as Slater determinants to allow for electron indistinguishability) in order to be valid for many-electron m.o.'s.

These many-electron m.o.'s will not normally allow the electrons to have a sharply defined total orbital angular momentum. They will not usually be eigenfunctions of $L^2_{(op)}$, etc., either singly or in combination, and so L^2, etc., will (usually) remain completely undefined, with an average value of zero. The presence of the nuclear framework is said to 'quench' the orbital angular momentum, for the interactions of the electrons with the various non-central nuclear force fields will change the direction of their motion continually, and thus give a fluctuating rather than constant L^2.

This should not be unexpected. The axes of the molecule are defined by the nuclear framework, so that if the electrons that bind this framework were to have any fixed non-zero orbital angular momentum, the whole molecule would behave like a gyroscope. It would only be necessary to find the magnetic field direction (relative to the nuclear framework) that did not lead to gyroscopic precession in order to discover the direction of L^2. The only exception to the rule that L^2 and L_z are undefined in molecules is in linear molecules and ions, where the nuclear framework only defines one molecular axis, and in highly symmetrical molecules and ions (e.g. benzene, $FeCl_6^{3-}$) where there is sufficient remaining ambiguity of molecular orientation to permit the possibility of at least partially defined orbital angular momentum. In the case of linear molecules, where the molecular axis is taken as being the z-direction, L_z is measurable, and for a given degenerate set of orbitals will have possible values of $\pm \Lambda h$, where Λ is a single integer which is determined by the l-values of the atomic orbitals from which the degenerate molecular orbitals are constructed. The fixed value of L_z is possible because of the unimpeded electron rotation that occurs about the molecular axis. L^2 will, however, remain undefined. The electron states of linear and highly symmetrical molecules are therefore complicated by the degeneracy considerations that we have dealt with for atoms, and have to be calculated analogously. They are called Σ, Π, Δ, etc., according to their Λ-values, by a rather loose

[13] See, for example, C. A. Coulson, *Valence*, 2nd edn, Oxford University Press, London, 1961, ch. 4.

correspondence with S, P, D, etc., atomic states. Similar terms are sometimes used more generally to describe some localized m.o.'s of types that would allow L_z to be defined if they were not in fact parts of larger molecules.

The above considerations restricting orbital motion do not apply to spin motion in the absence of spin-orbit coupling, because such spin motion is not in any way linked to the nuclear framework: the nuclear force fields cannot in general affect the electron spin orientation. Therefore in most molecules with unpaired electrons spin angular momentum is not 'quenched', and usually is the main determiner of the molecular paramagnetic susceptibility. The only exceptions to this occur when heavy atoms are present, for these show rather strong spin-orbit coupling; as a result, molecules with heavy atoms are often less paramagnetic than simple theory predicts.

Let us now return to larger molecules with simple, non-degenerate molecular orbitals. The spectra of such molecules are often quite easy to interpret. For example, in saturated organic ketones (e.g. acetone, $(CH_3)_2CO$) it is a fairly good approximation to say that the eight bonding electrons in the CO region are in localized m.o.'s as follows:

2 in σ, a C—O σ-bonding orbital
2 in π, a C—O π-bonding orbital
4 in n, non-bonding lone pairs on the oxygen

These are, reading downwards, in order of increasing energy. In addition, the following low-lying unoccupied orbitals are predicted, the order being as above:

π^*, an antibonding C—O orbital
σ^*, an antibonding C—O orbital

The actual spectra of such ketones confirm the validity of this generalization. They all show a single lowest wavenumber absorption in the range $33\,000$–$36\,500$ cm^{-1}, then one in the range $50\,000$–$52\,500$ cm^{-1}, and then one in the range $57\,000$–$60\,500$ cm^{-1}. Beyond this, as with almost all ultraviolet spectra, the molecules absorb almost continuously, as the wavenumber of the radiation is large enough for a wide variety of transitions to be possible, including those to states with higher principal quantum number ('Rydberg states'). The three lowest wavenumber peaks have been identified as due to $\pi^* \leftarrow n$, $\sigma^* \leftarrow n$, and $\pi^* \leftarrow \pi$ transitions.

The spectra of more complex molecules may often be interpreted by reference to simpler ones. Thus the spectrum of acetic acid, $CH_3CO.OH$, is similar to that of acetone except that (notably) the longest wavelength absorption is at $49\,100$ cm^{-1}. Now the n orbital is not likely to be greatly affected by substitution of —OH for —CH$_3$, and therefore what probably occurs is that the π^* orbital becomes less able to accept an electron because this involves strong repulsion of that electron by the p electrons on the hydroxyl oxygen. Substituents that shift the positions of spectral absorptions but do not change the general character of a spectrum are called chromophores. If the shift they induce is to lower wavenumber it is called a bathochomic or red shift, and if to higher wavenumber, hypso-chromic or blue.

A somewhat more complex interaction occurs in a conjugated com-pound such as mesityl oxide $(CH_3)_2C = CH.CO.CH_3$. In this com-pound it is not even approximately valid to treat the C=O π-electrons as localized. Instead, we must consider delocalized molecular orbitals spanning four atoms. The delocalized orbitals turn out to be two bonding (filled) and two antibonding orbitals (empty), each separated from the next by a smaller energy gap than the $\pi^* \leftarrow \pi$ gap in the individual bonds.

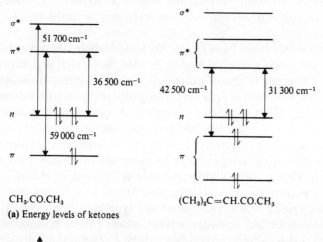

CH₃.CO.CH₃ $(CH_3)_2C=CH.CO.CH_3$
(a) Energy levels of ketones

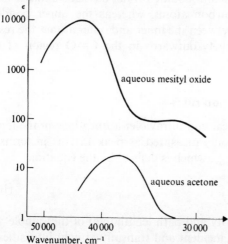

(b) Ultraviolet absorption spectra of aqueous ketones. (Reproduced in part with permission from *J. Amer. Chem. Soc.*, 1957, **79**, 1035. Copyright (1957) by the American Chemical Society.)

Fig. **10.3**

The lowest transition observed is now at $31\,300\,cm^{-1}$, and corresponds to a transition from n to the lower antibonding delocalized orbital; the next transition is even more markedly lowered in energy, at $42\,500\,cm^{-1}$, and corresponds to a transition from the highest occupied delocalized orbital to the lowest unoccupied one (see figure 10.3).

10.8 Solvent effects

All the wavenumbers quoted in the last section were approximate, for they depend quite noticeably on the solvent that is used if the spectrum is observed in the liquid state. For example, the $\pi^* \leftarrow n$ transition in acetone occurs at $35\,900\,cm^{-1}$ in hexane, but at $37\,800\,cm^{-1}$ in water. The ordering of solvents in between these extremes is in the order of their polarity.

The reason for this is not hard to see. An acetone molecule in hexane does not interact much with surrounding hexane molecules, and hence its absorption frequency is almost exactly that of the vapour. However, an acetone molecule in water experiences strong dipole-dipole interactions with nearby water molecules, and is strongly hydrogen bonded to some of them. After an $\pi^* \leftarrow n$ transition, however, the acetone molecule is not nearly so polar, because an electron has been moved from being approximately on one side of the oxygen atom to being approximately on the other side of it. Hence the ground state only is lowered in energy by solution, and therefore the $\pi^* \leftarrow n$ transition moves to a higher wavenumber. It is interesting to note that the $\pi^* \leftarrow \pi$ type transition in mesityl oxide shows a smaller but opposite solvent effect. This is presumably because the lower delocalized antibonding orbital has substantial density on the oxygen and outer carbon atoms, whereas the upper bonding orbital has its main density between the inner and outer atoms, the result being a shift of electron density outwards in the C=O region of the molecule upon excitation.

10.9 Intensities and selection rules

The intensities of broad optical and ultraviolet absorptions in the liquid (and solid) states are commonly measured as peak heights in terms of their extinction coefficient, ε_{max}, which is defined by the equation

$$\varepsilon_{max} = \frac{1}{Cl} \log_{10} \frac{I_0}{I} \tag{10.4}$$

where C and l are the molarity and path length (cm) of the sample and I_0 and I are respectively the incident and transmitted light intensities at the absorption frequency of the peak concerned. Extinction coefficients are an approximate measure of peak area, in so far as in most condensed phase optical spectra the linewidths of most peaks are similar. ε_{max} varies from less than 1 for a very weak absorption to almost $10^6\,l\,mol^{-1}cm^{-1}$, and its order of magnitude is a useful guide in identifying different types of absorption.

The total peak area of an optical absorption is determined by equation 4.19, generalized to three dimensions. The transition moments of electron transitions can be fairly accurately estimated for atoms but only rather approximately for molecules. The three underlying principles are as follows:

a *In most atoms and molecules no change in electron spin angular momentum is to be expected due to interaction with the electric field of radiation.*

This was discussed in section 4.7. In the absence of spin-orbit coupling there is no connection between spin motion and orbital motion, and only the latter is affected by an electric field. Therefore transitions that would involve a change of spin momentum direction are strongly forbidden—so much so that many atoms in which, for example, both singlet and triplet spin states are found exhibit spectra that appear to come from a 1:3 mixture of completely different substances. Even when spin and orbital motion are linked by weak spin-orbit coupling, the low probability of spin-change transitions is apparent in the phenomenon of phosphorescence (section 10.12(ii)). However, the rule breaks down in heavy atoms such as Hg, where spin-orbit coupling is large. For example, the strong $39\,400\,\text{cm}^{-1}$ mercury line is due to a $^3P_1 - {}^1S_0$ transition.

b *Angular momentum is conserved in the absorption and emission of photons.*

The absorption or emission of a photon by a system involves the addition to or subtraction from that system of a quantity of angular momentum $\sqrt{2}\hbar$, with a direction such that its component in the radiation direction is $\pm\hbar$. The consequences of this have already been discussed for molecules in section 9.9. Similar consequences hold for atoms, in which transitions are necessarily between exact angular momentum states.

In the case of one-electron atoms where changes of spin can be neglected, the consequences are that $\Delta l = \pm 1$ and $\Delta m = \pm 1$ or 0. There are several lines of argument by which these rules may be deduced, both for this problem and for the similar problem of the rigid rotor (chapter 8). We may illustrate one of these arguments[14] by considering the effect of absorption of a photon on a one-electron atom with $l = 2$ in a magnetic field parallel to the radiation direction (z). The electron has five possible initial z-components of angular momentum: $2\hbar$, $\hbar$, 0, $-\hbar$, and $-2\hbar$. The photon either adds or subtracts a z-component of $\hbar$, depending on the sense of its circular polarization. If we assume that a different state results in each case, we thus reach ten possible angular momentum states, with z-components: $3\hbar$, $2\hbar$, $\hbar$(twice), 0(twice), $-\hbar$(twice), $-2\hbar$, $-3\hbar$. On this

[14] See also problem 10.8.

basis alone we could very reasonably conclude, knowing that l has to be integral and that m has to go in integral steps from $+l$ to $-l$, that l had changed either to 3 (giving z-components of $3h$ to $-3h$) or to 1 (giving m states of h to $-h$). We could confirm this conclusion if we had knowledge of the appropriate operator for total orbital and photon angular momentum. Thus in this case we would have confirmed that $\Delta l = \pm 1$ and also proved the selection rule $\Delta m = \pm 1$. Transitions with $\Delta m = 0$ can arise, however, when the magnetic field direction and the radiation propagation direction are not parallel, although it still remains true that l cannot change other than by ± 1. A further analysis can be made to show that the $\Delta m = 0$ transitions occur only when the electric vector of the polarized radiation is parallel to the magnetic field, and the $\Delta m = \pm 1$ transitions occur only when they are mutually perpendicular.

With polyelectronic atoms and significant spin-orbit coupling there are many more ways in which a photon can change the angular momentum state. However, it still remains true that the addition of a photon momentum of only $\sqrt{2}\hbar$ with z-component $\pm \hbar$ cannot possibly change either J or m_J by more than unity, i.e.

$$\Delta J = \pm 1 \text{ or } 0, \quad \Delta m_J = \pm 1 \text{ or } 0$$

This may be simply proved, because $\sqrt{(J+2)(J+3)} - \sqrt{J(J+1)} > \sqrt{2}$ for all $J \geqslant 0$. The same selection rules as for Δm apply for Δm_J.

c *The transition moment must be non-zero.*

This requirement overlaps with requirement (b), but is more general. Using the independent electron approximation, the[15] total probability of a particular transition for random radiation direction (or for one radiation direction and random molecular orientation) is proportional to

$$\frac{1}{3}\left[\left(\int \psi_k^* x \psi_0 \, d\tau\right)^2 + \left(\int \psi_k^* y \psi_0 \, d\tau\right)^2 + \left(\int \psi_k^* z \psi_0 \, d\tau\right)^2\right]$$

where x, y, and z are the coordinates of the electron undergoing the transition. Some or all of these components may be zero for particular transitions in molecules possessing symmetry.

The ketone $\pi^* \leftarrow n$ transition mentioned previously provides an example of this. The n orbital is approximately a p_y atomic orbital centred on the oxygen atom, and the π^* orbital is in the xz plane (see figure 10.4).

The transition moment component

$$\int \psi_{\pi^*}^* x \psi_n \, d\tau \equiv \int_{-\infty}^{\infty} \int_{-\infty}^{\infty} \int_{-\infty}^{\infty} \psi_{\pi^*}^* x \psi_n \, dx \, dy \, dz$$

will be zero because in the integration over y the ψ_n orbital (and only the ψ_n orbital) has equal and opposite signs on either side of the xz plane. The same will hold true of the component involving z. Also, the com-

---

[15] Selection rules may be predicted without making this approximation. The reasoning, although formally more complex, is basically similar to that given here.

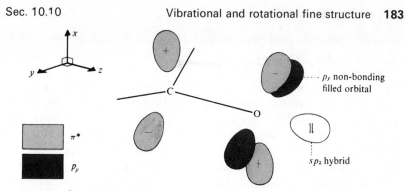

Fig. 10.4 $\pi^* \leftarrow n$ transition in ketones

ponents involving y will be zero for a different reason. y has equal and opposite sign on either side of the xz plane (i.e. is antisymmetrical about it), so that the product $y\psi_n$ is symmetrical about the xz plane. Hence the integration over y will not give zero. However, the integration over x will, because of the antisymmetry of the π^* orbital about the yz plane.

Thus it would appear that the $\pi^* \leftarrow n$ transition is totally forbidden. In fact it is observed, but with $\varepsilon_{max} \approx 20$ compared with $\varepsilon_{max} \approx 10^4$ to 10^5 for fully allowed transitions. The semi-classical explanation of this is that the molecule will frequently be slightly distorted due to vibrations; some of these distortions will distort the symmetry of the p_y and π^* molecular orbitals so that the transition moment integrals are no longer over exactly equal and opposite halves. A more rigorous quantum-mechanical treatment of this 'vibronic coupling' shows that an actual vibrational transition in a normal mode of appropriate symmetry must take place simultaneously with the electron transition in order for the latter to become weakly allowed (see problem 10.14).

A further rather obvious consequence of the transition moment integral is that ψ_k and ψ_C must be simultaneously substantial in the same region of space if the transition is to be strong. Many molecular orbitals are relatively localized in space, and therefore transitions between such localized orbitals are only observed when these are in the same part of the molecule, or over a very short distance to a neighbouring atom, ion, or molecule. Transitions of this latter intermolecular type form a special class known as 'charge transfer', an example being the intense colour of crystalline $CuBr_2$ due to the charge transfer transition $Cu^{2+} \leftarrow Br^-$. They occur typically with elements such as transition metals and heavy halogens which have a variety of readily accessible valence states.

10.10 Vibrational and rotational fine structure

The complete wavefunction of a molecule in a particular state, according to the Born-Oppenheimer approximation theory, consists of an electronic wavefunction ψ^e multiplied by an overall nuclear wavefunction ψ^n as discussed in previous chapters. When a spectroscopic transition occurs, both ψ^e and ψ^n will in general change, and as the changes in ψ^n will generally affect the total energy much less than those in ψ^e, they will be observable as fine structure of the electron absorption or emission spectrum bands, provided that the resolution is adequate.

The resolution will only be fully adequate, in fact, in gases and vapours at low pressure, and even then the fine structure will only be capable of analysis if the molecule is a relatively small one. However, this limitation is more than compensated for by the wealth of structural information that can be extracted from a well-resolved spectrum of a small molecule, including information that cannot be extracted from infrared spectra because of the stricter selection rules that apply to changes in ψ^n alone. Mention has already been made of $\Delta J = 0$ rotational transitions becoming possible when the photon angular momentum is instead taken up by the electron transition, but perhaps the most important restrictions to disappear are those limiting the allowed vibrational transitions. The individual nuclear vibrational wavefunctions in any one electronic state will not in general have the same centres of symmetry or extents as in some other electronic state, because every different electronic state of a molecule has its own equilibrium bond lengths, angles, and force constants. Thus transitions with any change of vibrational quantum number are allowed.

The relative intensities of these transitions may be calculated using the Franck-Condon principle. This useful approximation derives from the Born-Oppenheimer approximation, and postulates that during the time of an electron transition in a molecule, the nuclei do not move significantly. The way in which this can be used to predict the intensities of vibrational fine structure components is indirectly discussed in appendix A, in the context of Raman spectroscopy. The principle may be conveniently but somewhat classically visualized in the context of diagrams of energy against internuclear separation such as in figure 10.5. It

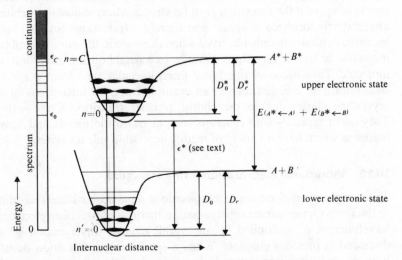

Fig. **10.5** Calculation of dissociation energies from vibrational fine structure

requires that transitions occur at constant internuclear separation and are hence represented by vertical lines in the diagram linking the high density parts of different $\psi^e \psi^n$ product states.

Dissociation energies

In pure infrared spectroscopy at normal temperatures only $n = 0$ to $n = 1$ transitions are generally observed, because of the selection rule $\Delta n = \pm 1$. As this rule does not apply to simultaneous electronic and nuclear transitions these reveal the energies of a much wider range of vibrational states, and hence permit a more comprehensive observation of anharmonicity effects (section 9.8). In particular it is often possible to calculate the dissociation energies of diatomic molecules both in the ground and excited electronic states from a knowledge of the vibrational levels in the excited states and of the spectra of the separated atoms. This is illustrated for a particularly simple case in figure 10.5, the strongest absorptive transitions being marked with vertical lines. The resulting spectrum is indicated on the left of the figure; it has been calculated on the assumption that only absorptive transitions from the $n' = 0$ vibrational level of the ground electronic state are important. This is valid at normal temperatures, although as the temperatures are increased new and separate series of weaker lines will appear, due to progressive thermal population of higher vibrational levels.

In this simple case one directly measures ε_C, the energy where transitions from the lowest level converge to a continuum. From this one can immediately deduce $D_0 = \varepsilon_C - E_{(A^* \leftarrow A)} - E_{(B^* \leftarrow B)}$. Also one can measure ε_0 and hence find that the dissociation energy in the excited state, $D_0^* = \varepsilon_C - \varepsilon_0$. The only real problem is that of identifying the excited atomic states A* and B*. However, this is usually not difficult provided that D_0 is known approximately from some separate source (e.g. an equilibrium constant). If the wrong excited states are guessed, the D_0 that one calculates will usually be a long way from the approximate value, whereas if the right states are guessed they will give a D_0 which is fully consistent with the approximate value and far more accurate.

Unfortunately it is not always possible to measure ε_C directly. In such cases it is necessary to extrapolate the converging spectrum from those lines that are observed. A rather approximate value of ε_C can be obtained by assuming that the convergence is regular. However, a much better method of finding ε_C exists, known as the Birge-Sponer extrapolation, which is valid even for rather irregular convergence. The energy ε_n of a transition from $n' = 0$ to level n of the upper state may be quite generally written $\varepsilon_n = \varepsilon^* + (n + \frac{1}{2})v_{\text{upper}} + p_n$, where p_n is a correction, almost always negative, for anharmonicity, and $\varepsilon^* = D_0 + E_{(A^* \leftarrow A)} + E_{(B^* \leftarrow B)} - D_e^*$. Hence

$$\varepsilon_0 = \varepsilon^* + \tfrac{1}{2}v_{\text{upper}} + p_0 \quad \text{and} \quad \varepsilon_C = \varepsilon^* + (C + \tfrac{1}{2})v_{\text{upper}} + p_C$$

Therefore

$$\varepsilon_C - \varepsilon_0 = Cv_{\text{upper}} + p_C - p_0$$

Also

$$\Delta\varepsilon_n \equiv \varepsilon_n - \varepsilon_{n-1} = \upsilon + p_n - p_{n-1}$$

and therefore

$$\sum_{n=1}^{C} \Delta\varepsilon_n = C\upsilon_{upper} + p_C - p_0 = \varepsilon_C - \varepsilon_0$$

Now $\sum_{n=1}^{C} \Delta\varepsilon$ is to a very good approximation the area under the graph of $\Delta\varepsilon_n$ against n, measured from $n = 0$ to $n = C$. This graph may be plotted for all measured values of $\Delta\varepsilon_n$ and then extrapolated towards $\Delta\varepsilon_n = 0$, i.e. the convergence limit $n = C$ and then the area under it may be measured. The advantage of the Birge-Sponer method is that even if the extrapolation is unreliable, the resulting inaccurate area under the extrapolated part ($\Delta\varepsilon_n$ small) will be a small proportion of the whole area.

Rotational fine structure

The rotational fine structure of electronic spectra will be similar to that of vibrational spectra because it is governed by the same selection rules $\Delta J = \pm 1$ or (sometimes) 0. We have seen that the $\Delta J = 0$ possibility is more likely in electronic spectra because of the greater variety of ways in which the photon angular momentum can be absorbed. It is also generally true that the moments of inertia in electronically excited states differ far more than those in the ground state than do those of states which are only vibrationally excited, and therefore we find that P- and R-branches converge or diverge, and Q-branches diverge, much more rapidly in electronic spectra than in pure vibrational spectra. The direction and extent of the divergence reveals the magnitude and sign of the change in bond length upon excitation.

10.11 Predissociation

Excited states are normally either bonding, in which case the energies and spectrum are as in figure 10.5, or antibonding, in which case there is no minimum in the upper curve and the spectrum is a continuum. However, a more complex possibility, called predissociation, is observed when two excited states occur with similar energies and symmetries. This possibility is illustrated in figure 10.6 by approximate levels for the m⌐lecule S_2.

The effect of the gap (which may be very small) is to produce an extra continuum in the spectrum, with blurred lines near the continuum where there is a finite probability of dissociation due to the small fraction of the vibrational wavefunction lying beyond the potential maximum. The rather curious excited states 1 and 2 may be thought of as deriving from a normal bonding state ($3\Sigma_u^-$) and a normal antibonding state ($3\Pi_u$) which 'mix' at those values of R at which their energies are similar.[16] Weak

---

[16] See C. A. Coulson, *Valence*, Oxford University Press, London, 1961, p. 68 for a general discussion of this 'non-crossing rule', and see also problem 10.11.

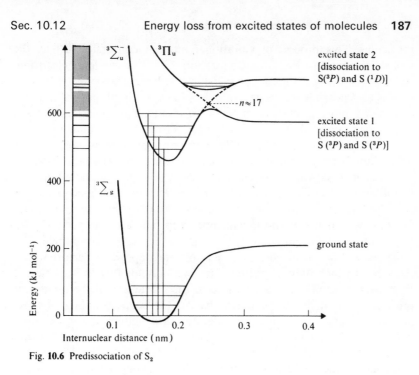

Fig. **10.6** Predissociation of S_2

mixing of this sort is quite common,[17] because molecular electronic states rarely have pure symmetry when all possible distortions are included. If the mixing is very weak, predissociation will not be observed in the absorption spectrum except, perhaps, as a slight broadening of lines near the cross-over point. However the mixing of the bonding and anti-bonding states will make molecular dissociation from the bonding state possible and the predissociation will be detectable by the unexpected appearance of free atoms. It is common parlance to describe the two states as if they were unmixed, and to describe the effect of their mixing as a 'radiationless transition'. However, such 'transitions' should not be confused with the various genuine transitions described in the next section, where the states are mixed by fluctuating rather than by static distortions.

10.12   Energy loss from excited states of molecules

It is very rare for a molecule in an excited vibrational state to lose its vibrational (and, where appropriate, electronic) excitation energy solely by radiation, although two cases where this does occur are mentioned below. Normally, the probability of spontaneous emission is much less (half-life $\approx 10^{-8}$ s) than that for vibrational internal conversion (half-life $\approx 10^{-12}$ s) in which the vibrational but not the electronic energy is lost either to other vibrational modes of the same molecule or to the motions of other

[17] See G. Herzberg, *The Spectra of Diatomic Molecules*, Van Nostrand/Reinhold, Princeton, 1950, p. 416 et seq., for a discussion of the symmetry requirements for strong predissociation.

molecules. This process of vibrational internal conversion brings the molecule rapidly to the ground vibrational level of whatever electronic state it is in. Further internal conversion involving changes of electronic state is also possible, but is generally less likely than vibrational internal conversion and thus has to compete with spontaneous emission. There is some uncertainty about the exact mechanism of internal conversion. Fortunately, however, we can understand a useful amount including the distinction between (rapid) vibrational internal conversion and (slow) internal conversion involving changes of electronic state, without knowing its full details. Let us suppose that a molecule is in an excited electronic state $*\psi^e$ and an excited vibrational state $*\psi^n$, so that its total wavefunction within the Born-Oppenheimer approximation is $*\psi^e*\psi^n$. We will assume for convenience that $*\psi^e$ and $*\psi^n$ are algebraically real. The molecule will be subjected to a variety of time-dependent perturbations, the sum of which may be written $P(t)$. We may identify at least two of these perturbations. One is that due to the vibrations of the other, not precisely normal modes of the molecule; the weak coupling of these modes with the mode under consideration will induce energy transfer, just as in vibrational spectra it induced combination bands. Another perturbation is that due to collision, in which both the nuclei and the electrons of the molecule experience the time-dependent Coulombic forces of other, moving nuclei and electrons as well as their own fixed mutual attractions and repulsions.

It is obvious that the first type of perturbation, $P_1(t)$, will fluctuate (when broken down into frequency components) at frequencies comparable to the energy level separations of the vibrationally excited modes. Each frequency component $P_1(v)$ of the variation of $P_1(t)$ with the normal coordinate q of the excited vibrational mode will also, in general, be analysable into a power series in q, i.e. $P_1(v) = {}_{i=0}\sum a_i q^i$, and therefore transitions, either stepwise or direct, to the ground vibrational level ψ^n of the excited electronic state $*\psi^e$ will be likely. For example, transitions between successive levels ψ_r^n and ψ_{r-1}^n will be strongly allowed because of the substantial transition moment $a_1\int\psi_r^n q\psi_{r-1}^n \, d\tau$. The only obvious exception to the above argument is a diatomic molecule, where there are no other vibrational modes available for internal conversion. In such molecules radiative fluorescence from $*\psi^e*\psi^n$ has indeed been observed, e.g. from excited N_2 molecules (nitrogen afterglow) and even from vibrationally but not electronically excited OH radicals produced chemically (i.e. from $\psi^e*\psi^n$).

The second type of perturbation, $P_2(t)$, due to collisions, will be important in higher pressure vapours and, of course, in condensed phases, Its frequency components will cut off quite sharply above a certain frequency (section 6.2(iii)) and, as this cut-off frequency is always lowered by slowing up the intermolecular motions, $P_2(t)$ will be less generally effective at lower temperatures in linking any but the closest energy levels. $P_2(t)$ is also particularly important in inducing changes of electronic state, which are largely unaffected by $P_1(t)$ because, to a first approximation, this only affects nuclear coordinates.

[$P_1(t)$ and $P_2(t)$ are of course not the only possible sources of relaxation of excited states. Experimentally, one observes that many low pressure gas phase samples have appreciable linewidths (i.e. relaxation) even for electron transitions only involving ground vibrational levels. The collision independent widths are found to be greatest when the molecule is large (i.e. has many vibrational modes) and also when the upper electronic state involved is fairly close in energy to other excited states. Under these circumstances the simple two-level perturbation theory of chapter 4 is not strictly valid, and it is not surprising to find these deviations from the two-level prediction of very narrow lines, in proportion to the number of further nearby levels. The full theory of linewidths in the presence of many near degenerate states is still in process of development. However, many observed phenomena, as in sub-section (ii) below, may be explained by approximately describing this breakdown of the Born-Oppenheimer approximation as being due to a new perturbation, $P_3(t)$.]

We will now consider in more detail the possible fates of a molecule which has, by the $P_1(t)$ mechanism described above, reached the lowest vibrational level of its excited electronic state. A range of possibilities are known experimentally, as illustrated in figure 10.7, and as described below.

(i)   Fluorescence directly from the upper state (radiative fluorescence)

The molecule may simply radiate by spontaneous emission from the ground vibrational but excited electronic state $*\psi^e \psi^n$, falling to whichever vibrational states $*\psi_n^n$ of the ground electronic state ψ^e give a substantial transition moment. Spontaneous emission corresponds in this instance to an interaction of the electrons with 'free space', and its probability is controlled by a transition moment $\int *\psi^e \psi^n . r_e . \psi^e *\psi^n d\tau$ very similar to that for an ordinary absorption. r_e is a shorthand for the x, y, or z coordinate of the electron, and does not affect the nuclear coordinate, and hence we may separate a factor $\int \psi^n . *\psi^n d\tau_{nuclei}$ from the integral. This is simply the overlap integral between ψ^n, the ground vibrational level of the upper electronic state and $*\psi^n$, which can be any vibrational level of the lower electronic state. There is thus an approximate mirror symmetry between the vibrational fine structure of ordinary absorption spectra ($*\psi^e *\psi^n \leftarrow \psi^e \psi^n$) and that of fluorescence emission spectra ($*\psi^e \psi^n \rightarrow \psi^e *\psi^n$) provided that the vibrational frequencies of the upper and lower electronic states are similar.[18]

The probability per molecule of such fluorescence at visible wavelengths is typically once in 10 seconds; this probability increases rapidly

[18] This is illustrated, along with a useful general discussion of energy transfer processes, in A. Cox and T. J. Kemp, *Introductory Photochemistry*, McGraw-Hill, Maidenhead, 1970, chs 1 and 2.

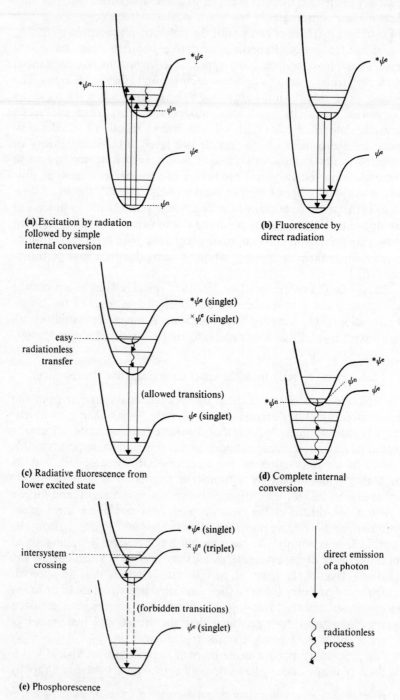

(a) Excitation by radiation followed by simple internal conversion

(b) Fluorescence by direct radiation

(c) Radiative fluorescence from lower excited state

(d) Complete internal conversion

(e) Phosphorescence

Fig. 10.7 Energy loss from excited molecular states

with increasing frequency. It means re-emission at a lower frequency than that of the initial exciting radiation, and this together with the next process below explains the usefulness of fluorescent materials in converting ultraviolet to visible light in neon tubes, detergent whiteners, etc.

(ii)   Fluorescence from a lower excited state

The above process (i) normally occurs only when $*\psi^e$ is the lowest excited electronic state. If there is an intermediate electronic state $\times\psi^e$ then in almost all cases (azulene being an exception) internal conversion occurs to this state. This illustrates a general rule, also true for ground states which is that internal conversion to any electronic state which is reasonably close in energy to $*\psi^e$ occurs in preference to direct emission to the ground state. There are at least two possible mechanisms for this internal conversion. One is a transition caused by $P_2(t)$ from $*\psi^e\psi^n$ to $\times\psi^e*\psi^n$, where $*\psi^n$ is the upper vibrational level of the state $\times\psi^e$ closest in energy to $*\psi^e\psi^n$. Such a transition involves a small energy jump, which is made possible by lattice vibrations and collisions in condensed phases at normal temperatures. It is also made possible in a lattice-independent way by the breakdown of the Born-Oppenheimer principle mentioned earlier in this section. This breakdown may be conveniently but approximately described as a perturbation $P_3(t)$ (the difference between the correct and the approximate Hamiltonians) which in accord with equation 10.2 has a considerable mixing effect on nearly degenerate levels. The degree of mixing is determined by integrals of type $\int(\times\psi^e*\psi^n)P_{2\text{ or }3}(t)(*\psi^e\psi^n)\,d\tau$, which will generally be substantial only if $*\psi^n$ is not too highly excited a vibrational level. For if $*\psi^n$ is highly excited it will oscillate many times on either side of zero over the extent of ψ^n, and thus the integral of the product $(*\psi^n)(\psi^n)$ will be rather small.[19] This explains the previously quoted observation that internal conversion direct to the electronic ground state is less likely in most molecules than internal conversion to a lower electronically excited state. It is quite common, in fact, to find a molecule which proceeds purely by internal conversion from $*\psi^e*\psi^n$ through to $\times\psi^e\psi^n$, and then fluoresces by direct radiation to some vibrational level of the ground electronic state ψ^e (see figure 10.7). Such a process will obviously produce a more dramatic wavelength change in radiative fluorescence than will direct radiation from $*\psi^e\psi^n$.

Phosphorescence

A special case of such internal conversion, called intersystem crossing, occurs when the lower excited state has a different spin multiplicity from the upper one. Typically, intersystem crossing can occur between an excited singlet state and a slightly less excited triplet state, provided that

---

[19] This is true even if ψ^n is replaced by $P_{2\text{ or }3}(t)\psi^n$, for neither is likely to oscillate sharply about zero as the nuclear coordinate is varied.

$P_2(t)$ affects the spin as well as the space coordinates. This is particularly likely when the perturbing collision involves a heavy atom, whose spin-orbit coupling is large; experimentally, the substitution of a xenon for an argon trapping matrix has been shown to increase the probability of one of the singlet-triplet transitions in benzene by a factor of 200. Alternatively, a heavy atom within the excited molecule itself can lead to the singlet-triplet transition by reducing the separability of the spin parts of $^\times\psi^e$ and $^*\psi^e$ from their space parts.

Once the system is in a triplet excited state, its direct radiation to the ground state will be extremely slow because of the $\Delta S = 0$ selection rule discussed previously. Emissive transitions taking, on average, more than 10^{-6} seconds to occur are known as phosphorescent transitions (with the exception of a few transitions with non-exponentially decaying emission in the 10^{-4} to 10^{-6} second range which are technically known as delayed fluorescence transitions, and arise from kinetically controlled inter-molecular interaction of two molecules in triplet states to give an excited singlet state which fluoresces normally). The applications of phosphorescence are well known in the form, for example, of light switches which glow for several minutes after the light is extinguished.

(iii)   Complete internal conversion

The probability of stepwise internal conversion from $^*\psi^e {}^*\psi^n$ all the way to $\psi^e \psi^n$ was not ruled out in the previous section, particularly when $^*\psi^e$ was not a very highly excited state, and when violent collisions were probable. This process (also technically known as fluorescence) does occur in many instances, particularly in mobile samples where there are many lattice motions of sufficient rapidity to allow transitions between quite widely spaced vibrational levels. The increased possibility of complete internal conversion explains why radiative fluorescence is less common and phosphorescence almost unknown in mobile liquids. (An interesting exception to this is the recently observed phosphorescence of benzophenone in perfluoroalkanes, possibly due to the absence of solvent protons and their rapid motions.)

The various mechanisms described above are shown collected together in diagrammatic form in figure 10.7.

In addition to these processes, long-range energy transfer is often possible in condensed phases. A mechanism for this is discussed in section 10.14.

10.13   Luminescence spectroscopy

Phosphorescence and radiative fluorescence are collectively known as luminescence; when the excitation is by chemical reaction rather than by absorption of radiation the radiation emission is called chemilumin-escence. Most luminescence can be detected spectroscopically, and, because its frequency from any one molecule will be generally independent of the exciting frequency, its spectrum will be a rather approximate fingerprint for the luminescing molecule. This may be of great use in the study of complex mixtures of molecules, as in biochemistry; for example, the strong luminescence of riboflavin (vitamin B_2) at 530 nm is a valuable means for identifying it. Similarly the different bioluminescence peaks of fireflies (562 nm) and *photobacterium fischeri* (490 nm) indicate different chemiluminescence mechanisms.

10.14   Lasers

It has already been suggested that direct radiative fluorescence from an excited state may be greatly enhanced by stimulated emission. This is best understood with reference to a rather simple system—atomic caesium vapour. The energy levels relevant to our present discussion are shown in figure 10.8. The 8p ← 6s transition at approximately $25\,700\,\mathrm{cm}^{-1}$ may be very conveniently and powerfully stimulated by surrounding the Cs vapour with a helium flash tube, which emits strongly in this region, and this radiation is sufficiently intense to saturate the transition briefly despite the presence of the normal de-excitation mechanisms. One reason for the success of the saturation is that any $25\,700\,\mathrm{cm}^{-1}$ radiation that is re-emitted may well be re-absorbed by other Cs atoms still in the 6s state; this is known as self-trapping of radiation.

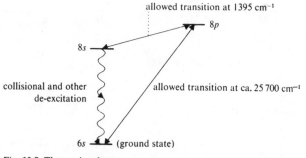

Fig. **10.8** The caesium laser

Of course, the 8p Cs atoms may well spontaneously emit small amounts of radiation at other wavenumbers, notably at the infrared wavenumber $1395\,\mathrm{cm}^{-1}$, which brings the atom to the 8s level. But such a small amount of radiation will stimulate the coherent emission of further $1395\,\mathrm{cm}^{-1}$ radiation (section 4.8), and this will stimulate still more coherent radiation particularly if this is partially reflected back by mirrors, so that a large pulse of coherent $1395\,\mathrm{cm}^{-1}$ radiation will be rapidly emitted. This process is called laser emission, and occurs in a wide variety of systems. The strength of the laser pulse makes it a valuable source of extremely high, although frequently intermittent radiation intensity, useful in fundamental research and in optical surgery, and its coherence leads to extreme linearity of the beam (useful for communication over long distances) and to interferometric applications (as in accurate measurement of long distances, e.g. earth to moon).

10.15   Absorption spectra of molecular crystals

Figure 10.9 shows parts of the absorption spectra of bromobenzene (in diagrammatic form) in its three possible phases.

In the vapour phase spectrum the lines are very sharp, and some rotational fine structure is resolved. In the solution spectrum the lines are much broader, although the vibrational structure is still discernible. Also, there is a small red shift in the main absorption frequency, presumably due to preferential stabilization of the excited state by solvation. In

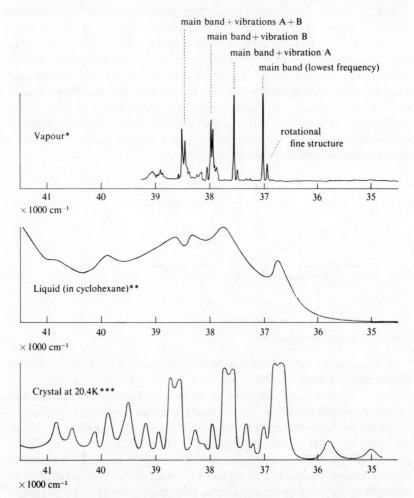

Fig. **10.9** Absorption spectra of bromobenzene. *From W. W. Robertson and F. A. Matsen, *JACS*, 1950, **72**, 5253; **from W. F. Forbes, *Canad. J. Chem.*, 1961, **39**, 1133; ***from A. S. Davydov, *Theory of Molecular Excitons*, McGraw-Hill, New York, 1962, p. 139.

the very low temperature single crystal spectrum the lines are much narrower than in the liquid, although not as narrow as in the vapour. However, new lines appear in the solid which are somewhat similar to, but more complex than the gas phase vibrational fine structure. Also, a study using polarized radiation reveals that many of the absorptions are strongly anisotropic; their anisotropy is greater than that which can be explained from the properties of individual non-interacting molecules, for these are not all parallel in the crystal. We are thus led to interpret the solid state absorption spectrum as arising from the electron motions not just of isolated molecules but of interacting molecules.

The simplest way of beginning to analyse solid state spectra is to use the classical model outlined in chapter 3. Consider first two parallel dipole oscillators in which only the negative charges move (fig. 10.10a).

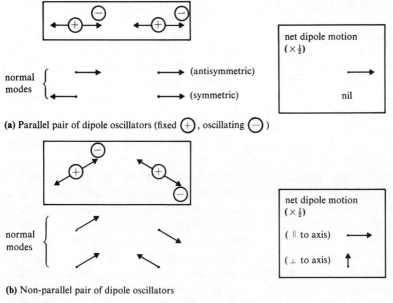

(a) Parallel pair of dipole oscillators (fixed (+), oscillating (−))

(b) Non-parallel pair of dipole oscillators

Fig. **10.10**

Two normal modes will exist, as shown in the figure. The antisymmetric mode will not involve a change in the mutual separation of the negative charges, and therefore any separation-dependent force between these will not affect their motion; the frequency will be unaltered and the motion will involve a net fluctuating dipole moment which will thus make ordinary light absorption possible. The symmetric mode, however, will have a different frequency because of the interaction between the negative charges. A simple repulsion between then, for example, will enhance the combined frequency by effectively increasing the force constants of the original oscillators. It would thus appear that the optical absorption will be split—until we remember that the symmetric mode involves no net fluctuating dipole and is hence spectrally inactive.

However, neighbouring molecular dipole oscillators will not generally be parallel. Indeed, within one unit cell they will almost always not be parallel. Figure 10.10(b) illustrates the non-parallel case for two oscillators, and predicts two absorptions, rather as before except that the second one is active to radiation polarized at right angles to the main axis. In general, a unit cell with n non-parallel molecules will split a given absorption of the isolated molecule into n sub-bands with different polarizations, each sub-band being somewhat broadened by further more distant interactions.

A real crystal is further complicated by molecular vibrations and other thermal motions. However the complete classical model gives fairly reliable qualitative predictions, and may also be applied, in modified form, to the pure vibrational spectra of molecular solids. In quantum mechanism the 'couplings' are identified as forces of Van der Waals and Coulomb repulsion type.

10.16 Excitons

So far we have only discussed normal modes of motion, as are observed spectroscopically. But now let us suppose that one oscillator is selectively excited, perhaps due to absorption at a different frequency followed by loss of excited state vibrational energy. This oscillator will, in classical terms, force oscillations of equal frequency, but lagging 90° in phase (section 3.3) in neighbouring oscillators because of the couplings between the oscillators. In the process it will lose its own energy, because the neighbouring oscillators will begin to drive it 180° out of phase with its original motion. The oscillation energy will thus spread outwards in an uneven ripple of energy. In a quantum system the average energy spread of the individual excitation will be (qualitatively) the same as in the classical system, but, like a free photon, it will appear to be localized if a measurement is made that determines (and thus affects) its position (see section 1.10). The excitation will thus move through the crystal like a quantum energy particle, its average position being guided by laws of quantum vibrational wave motion. For this reason optical energy excitations in crystals are called 'excitons', and vibrational excitations 'phonons', by analogy with photons. The rate of motion of excitons may, of course, be slow, if the couplings are weak. In this case a localized crystal distortion can develop around the exciton, further slowing its motion and making the absorption spectrum of the solid similar to that of the liquid. But the exciton may move rapidly, so that energy is rapidly transferred through the crystal without degradation to vibrational motions. This rather rapid energy transfer is important in current theories of biological photochemistry, and is also observed as a long-range process of inter-molecular energy transfer in photochemical studies. It is not necessary for the coupled oscillators to be the same molecule provided they have nearby absorption energies, and some molecules (e.g. benzophenone) are well known as phosphorescence sensitizers for other molecules (e.g. naphthalene) due to their ability to absorb radiation, cross to a triplet state, and then transfer an exciton of suitable energy to the phosphorescing molecule, typically over a distance of up to 1.5 nm. It is necessary for the separate excitations to involve the same spin change in order for them to couple strongly.

Suggestions for further reading

H. H. Jaffé and M. Orchin, *Theory and Applications of Ultraviolet Spectroscopy*, John Wiley, New York, 1962.

C. N. R. Rao, *Ultraviolet and Visible Spectroscopy*, Butterworths, London, 1967.

G. Herzberg, op. cit., vol. III.

L. Pauling and E. B. Wilson, *Introduction to Quantum Mechanics*, McGraw-Hill, New York, 1935.

H. C. Kuhn, *Atomic Spectra*, Longmans Green, London, 1962.

A. Cox and T. J. Kemp, *Introductory Photochemistry*, McGraw-Hill, Maidenhead, 1970.

D. S. McClure, *Electronic Spectra of Molecules and Ions in Crystals*, Academic Press, New York, 1959.

Problems

10.1 The angular parts of the $3d$ orbital wavefunctions (which all have the same radial part) are

orbital	angular part of ψ
d_{z^2}	$3\cos^2\theta - 1$
d_{xy}	$\sin^2\theta \sin 2\phi$
d_{yz}	$\sin\theta \cos\theta \sin\phi$
d_{xz}	$\sin\theta \cos\theta \cos\phi$
$d_{x^2-y^2}$	$\sin^2\theta \cos 2\phi$

Show that those orbitals which transform into each other upon rotation about the z-axis combine to give eigenfunctions of the operator $-i\hbar\,\partial/\partial\phi$, and find the eigenvalues of the combinations.

10.2 Show that the first-order perturbation energy of an atom due to a static applied electric field E_0 (Stark effect) will be zero, whatever eigenstate the unperturbed atom is in (see also problem 4.4).

   Why is this proof not valid for the unique case of the hydrogen atom, except for the ground state? Find an expression for the first-order perturbation energies of the combinations $\psi_{2s} \pm \psi_{2p(z)}$ for an electric field in the z-direction. What are the perturbation energies of $\psi_{2p(x)}$ and $\psi_{2p(y)}$?

10.3 Consider two H atoms a and b with wavefunctions ψ_a and ψ_b, brought together from a great distance.

   a Write down the two correct antisymmetrized wavefunctions ψ_c and $\psi_{c'}$, including spin, for the combined $H_a + H_b$ system, assuming no interaction between the atoms.

   b Show that these combinations satisfy the requirement that $\int \psi_{c'} P \psi_c = 0$, where P is the (weak) interaction between the atoms.

   c Hence calculate the perturbation energies of ψ_c and $\psi_{c'}$ in terms of integrals such as $\int \psi_a(1).\psi_b(2)P\psi_a(2).\psi_b(1)\,d\tau$.

   d Also calculate the energy of a spectroscopic transition between the two states, and the transition moment of this transition. What will be the direction of the transition moment?

10.4   Using the arguments of sections 10.4 and 10.5, explain why the ground configuration of the nitrogen atom, $1s^2 2s^2 2p^3$, does not contain any of the states ^4F, ^2F, ^4D, ^4P.

In fact, the configuration consists of ^2D, ^2P, ^4S. Use this information to justify the presence of the states ^3D, ^1D, ^3P, ^1P, ^5S, ^3S in the $1s^2 2s^1 2p^3$ excited configuration of carbon. Which of these states will be lowest?

10.5   The operator corresponding to the spin-orbit coupling energy for a single electron in an atom may be written as $\zeta(l_{(op)} \cdot s_{(op)})$, where $l_{(op)}$ and $s_{(op)}$ are the operators corresponding to the orbital and spin angular momenta respectively, $l_{(op)} \cdot s_{(op)}$ is a scalar product and ζ is a constant. Using the scalar product relation

$$l_{(op)} \cdot s_{(op)} = l_{x(op)} s_{x(op)} + l_{y(op)} s_{y(op)} + l_{z(op)} s_{z(op)}$$

show that the spin-orbit coupling operator commutes with $j^2_{(op)}$, the combined squared angular momentum operator, $(j_{(op)} = l_{(op)} + s_{(op)})$, and hence that j^2 may still be precisely measured in the presence of spin-orbit coupling.

10.6   State whether the following transitions are allowed or forbidden, and if they are allowed, give the possible directions of the transition moment;
  a $3d \leftarrow 2s$ and $3p \leftarrow 1s$ in the hydrogen atom,
  b $^1D \leftarrow ^1S$ and $^3P_1 \leftarrow ^3P_0$ in the $1s^2 2s^2 2p^2$ carbon atom,
  c $\pi^* \leftarrow \pi$ and $\sigma^* \leftarrow \pi$ in the CO molecule,
  d $e_g \leftarrow t_{2g}$ in a transition metal complex (e_g and t_{2g} are labels for metal d-electron levels perturbed by the electrostatic repulsions of the ligands in the complex).

10.7   In a recent experiment, bromine vapour was sealed in a glass sphere coated (except for two patches at opposite ends of a diameter) with a thin film of metallic tellurium. A beam of plane polarized light of frequency just above that needed to cause the dissociation $Br_2 \rightarrow 2Br$ was passed along the diameter through the clear patches. The observation (simplifying slightly) was that the Te film was removed, by the reaction $Te + 2Br \rightarrow TeBr_2$, in two broad patches above and below the plane of polarization of the light. Use this information, together with the fact that the atoms in a diatomic molecule can only dissociate along the internuclear axis, to deduce the direction of the transition moment of the dissociative transition, relative to the internuclear axis. What transition might this be?

10.8   The θ and ϕ part of the hydrogen-like wavefunction for quantum numbers l and m contains, as its terms with maximum power, $f^a(\theta) f^b(\phi)$, where $f^n(x) = \cos^n x$ or $\sin^n x$ or any product of $\cos x$ and $\sin x$ with total power n, and where $a = l$ and $b = |m|$. Use this fact, together with the argument concerning the vibrational selection rule in appendix 1, to deduce the selection rules $\Delta l = \pm 1$ and $\Delta m = \pm 1$ or 0 for electric dipole transitions.

10.9   The lowest frequency electronic transition in methyl chloride undergoes a blue shift when the solvent is made more polar. However, the frequency shifts to the red if bromine is substituted for chlorine. It shifts further still in methyl iodide, which is normally kept in darkened bottles because it is decomposed by daylight. It also shifts progressively to the red in the series CH_3Cl, CH_2Cl_2, $CHCl_3$, CCl_4. What explanation can you offer for these facts?

10.10 Explain the following experimental observations:

  a The light output of a domestic discharge tube is greatly improved by the fluorescent internal tube coating.

  b 1-bromonaphthalene has a phosphorescence lifetime of 0.018 s and is 164 times as likely to phosphoresce as to fluoresce (directly). The corresponding figures for 1-chloronaphthalene are 0.20 s and 5.2 times, whereas for naphthalene itself the figures are 2.3 s and 0.09 times.

  c The phosphorescence lifetime of perdeuterated polyacenes exceeds those of their ^{1}H analogues by a factor of between 2 and 8, the higher reductions being at lower temperatures.

  d Some weak predissociations become appreciably stronger as the pressure of the vapour under observation is increased.

10.11 An approximate m.o. calculation for a molecule gives various orthonormal orbitals ψ which are eigenfunctions of a simplified Hamiltonian H. Two of these, ψ_a and ψ_b, whose eigenvalues are E_a and E_b, are close in energy, and at a particular value of the internuclear distance (with which all the ψ vary) $E_a = E_b$. However, no other orbitals come close in energy to ψ_a and ψ_b.

  The various energy terms which were omitted from the simplified Hamiltonian are now considered as a perturbation P, so that the full Hamiltonian is $H + P$.

  a Use the argument of section 10.2 to show that for some value of λ, $\psi_a + \lambda\psi_b$ must be a correct (unnormalized) eigenfunction of $H + P$ at any particular value of the internuclear distance.

  b Using formula 1.12 show that the energy of $\psi_a + \lambda\psi_b$ is

$$\frac{(\lambda^2 - 1)(E'_a - E'_b)}{2(1 + \lambda^2)} + \frac{2\lambda}{1 + \lambda^2}\int\psi_a^*P\psi_b\,d\tau \quad \text{below} \quad \frac{E'_a + E'_b}{2}$$

  where E'_a and E'_b are equal to $E_a + \int\psi_a^*P\psi_a\,d\tau$ and $E_b + \int\psi_b^*P\psi_b\,d\tau$ respectively.

  c Show that this energy difference is a minimum when $\lambda/(\lambda^2 - 1) = \int\psi_a^*P\psi_b\,d\tau/(E'_a - E'_b)$, and hence that at this minimum the difference is given by

$$\frac{\lambda}{(1 + \lambda^2)}\left(\frac{(E'_a - E'_b)^2 + 4\left[\int\psi_a^*P\psi_b\,d\tau\right]^2}{2\int\psi_a^*P\psi_b\,d\tau}\right)$$

  Hence show that $(\psi_a + \lambda\psi_b)$ must always have an energy lower than $(E'_a + E'_b)/2$ if λ has the same sign as $\int\psi_a^*P\psi_b\,d\tau$ and greater if λ has the opposite sign, and hence that $(\psi_a + \lambda\psi_b)$ and its orthogonal pair $(\lambda\psi_a - \psi_b)$ will never cross in energy. (It may help to show that, if $(\lambda\psi_a - \psi_b)$ is rewritten $(\psi_a - (1/\lambda)\psi_b)$ then it also fits the relationship given at the beginning of this section.)

  d Show that if P is spherically symmetrical, then unless ψ_a and ψ_b have some of the same symmetry the levels will touch in energy, and that despite the result of **c** this will be equivalent to their crossing.

10.12 The following information holds true for a two-electron atom:

$$S^2_{(op)} = S^2_{1(op)} + S^2_{2(op)} + 2S_{1z(op)}S_{2z(op)} + S_{1+(op)}S_{2-(op)} + S_{1-(op)}S_{2+(op)}$$

$$S^2_{1(op)}\alpha(1) = \tfrac{3}{4}\hbar^2\alpha(1) \qquad\qquad S^2_{1(op)}\beta(1) = \tfrac{3}{4}\hbar^2\beta(1)$$

$$S_{1z(op)}\alpha(1) = \tfrac{1}{2}\hbar\alpha(1) \qquad\qquad S_{1z(op)}\beta(1) = -\tfrac{1}{2}\hbar\beta(1)$$

$$S_{1+(op)}\alpha(1) = 0 \qquad\qquad\qquad S_{1+(op)}\beta(1) = \hbar\alpha(1)$$

$$S_{1-(op)}\alpha(1) = \hbar\beta(1) \qquad\qquad S_{1-(op)}\beta(1) = 0$$

S_1 operators only affect electron (1), and S_2 operators electron (2).

Prove that ψ_p, ψ_q, $\psi_r + \psi_s$, and $\psi_r - \psi_s$ of section 10.4 are eigenfunctions of $S^2_{(op)}$ and of $S_{z(op)}$, and find their various eigenvalues.

10.13 Show that the two-electron density represented by the wavefunction $\psi_r - \psi_s$ (section 10.4) is substantial in the region where the 1s and 2s one-electron atomic orbitals overlap, but that conversely $\psi_r + \psi_s$ is rather smaller in this region. Hence explain why the triplet state of the helium $1s^1 2s^1$ configuration is of lower energy than the singlet state.

10.14 Consider the orbitals ψ_n and ψ_{n*} illustrated in figure 10.4. Show that $\int\psi_{n*}^* xy\psi_n\,d\tau \neq 0$, and that therefore $\pi^* \leftarrow n$ transitions may be induced by x- (or y-) polarized light provided that they are simultaneously accompanied by a change $\Delta n = \pm 1$ in a vibrational state whose normal coordinate varies in the region of the CO bond as does y (or x). It may be of help to consider the argument given in appendix 1 for the $\Delta n = \pm 1$ selection rule in vibrational spectroscopy.

10.15 The following emission bands (among others) were observed in the fluorescence spectrum of iodine vapour corresponding to transitions from one particular excited state of I_2 to the various vibrational levels n of the ground state.

n	$v(\text{cm}^{-1})$	n	$v(\text{cm}^{-1})$
0	54 619	36	47 826
4	53 778	40	47 195
6	53 365	43	46 741
8	52 956	46	46 305
10	52 552	50	45 752
13	51 958	55	45 111
17	51 183	58	44 755
21	50 431	62	44 318
26	49 523	66	43 924
30	48 825	71	43 479
33	48 318	76	43 147
		83	42 783

a Plot $(v_a - v_b)/(n_a - n_b)$ for successive lines against $\tfrac{1}{2}(n_a + n_b)$, and hence find the value of n for which $\Delta v = 0$.

b Use this information, together with the v values given, to deduce the chemical dissociation energy of I_2.

Appendix 1
Infrared and Raman selection rules

Infrared

The vibrational wavefunctions with quantum number n of the individual nuclei of mass m in a molecule are of the form

$$\psi_n = H_n(z) e^{-z^2/2} \tag{A1.1}$$

where $z^2 = \beta x^2$, $\beta = \sqrt{mk}/\hbar$, k is the bond force constant and $H_n(z)$ is the nth Hermite polynomial. Hermite polynomials are mathematical functions somewhat analogous to $\cos nz$, and the first three are $H_0(z) = 1$, $H_1(z) = 2z$, $H_2(z) = 4z^2 - 2$. $H_n(z)$ is always an even function, i.e. it does not change sign when $-z$ is substituted for z, provided n is even. Conversely, it is always an odd function if n is odd. The highest power of z in any $H_n(z)$ is always $(2z)^n$. Their derivation and properties are described in most quantum mechanics textbooks. One of the resulting wavefunctions is plotted in figure 9.2. An important property of the wavefunctions ψ_n is that any well-behaved one-variable function ψ in the range $-\infty$ to $+\infty$ can be written as a series in terms of them, i.e. $\psi = {_n}\sum a_n \psi_n$ (just as functions repeating in the range $-\pi$ to $+\pi$ can be described by a Fourier series). The ψ_n are also, of course, orthogonal.

The transition probability for an electric dipole induced vibrational transition from state n to state m is proportional to the square of the transition moment integral, which in this case has the form $\int \psi_m^* x \psi_n \, dx$. (The radiation field must be parallel to the direction of vibration and hence only the x component of $\vec{r}$ is relevant.)

Let us consider $\int \psi_m^* x \psi_n \, dx$ in more detail, substituting A1.1 for the ψs and z^2 for βx^2. Then the transition intensity is proportional to the square of

$$\frac{1}{\beta} \int H_m(z)^* z H_n(z) e^{-z^2} \, dz \tag{A1.2}$$

For the sake of argument let us assume that m is the upper state. Now $z H_n(z) e^{-z^2/2}$ is not a vibrational wavefunction but it may be written as a series expansion in terms of vibrational wavefunctions.

$$z H_n(z) e^{-z^2/2} = \sum_r a_r H_r(z) e^{-z^2/2}$$

Furthermore, r cannot exceed $n+1$, because the maximum power of z in any Hermite polynomial $H_r(z)$ is z^r. Clearly, multiplying $H_n(z)$ by z

201

cannot produce powers of z greater than $n+1$, and hence no contributions from Hermite polynomials higher than $H_{(n+1)}$ are needed to construct $zH_n(z)$. Now all vibrational wavefunctions ψ_n are orthogonal, and hence unless $m = n+1$, A1.2 will be identically zero because in no other way can $m = r$. Hence we have derived the selection rule $n = +1$ for absorption, and, by analogy, the corresponding rule $n = -1$ for emission.

Raman

The derivation of the corresponding rule for Raman spectroscopy is considerably more complex. The transition probability is proportional to the square of expression 9.6. An expression of this complexity is only amenable to approximate analysis, and the following analysis is due to Ting.[1]

Expression 9.6 may first be simplified by using the Franck-Condon principle, as discussed in section 10.10. In this approximation, the nuclei do not move during the entire time in which a photon undergoes Raman scattering. The Raman process is thought of as a double electron jump in which the electrons return to their original state but during their two transitions give the nuclei a mechanical impulse. This impulse changes the nuclear velocity, and hence vibrational state, but not (within the time-span of the electron jumps) the nuclear positions.

As before, we separate nuclear and electron states, so that

$$\psi_j = \psi_j^n \psi_j^e, \quad \psi_G = \psi_G^n \psi_G^e \quad \text{and} \quad \psi_k = \psi_k^n \psi_k^e$$

the superscripts n and e denoting nuclear and electronic respectively. Our model above implies that $\psi_j^e = \psi_G^e$, and therefore that the transition probability is proportional to the square of

$$\sum_k \frac{\int \psi_j^{n*} \psi_k^n \, d\tau \cdot \int \psi_k^{n*} \psi_G^n \, d\tau}{v_{jk} + v} \cdot \int \psi_G^{e*} e\vec{E}_0 \cdot \vec{r} \psi_k^e \, d\tau \cdot \int \psi_k^{e*} e\vec{E}_0 \cdot \vec{r} \psi_G^e \, d\tau \quad \text{(A1.3)}$$

The right-hand two integrals concern only the electronic transitions and merely put some restrictions on the actual upper states that are involved in the electric polarization. The left-hand two integrals are simply nuclear overlap integrals. One sees immediately that if the upper state has exactly the same vibrational frequency and equilibrium nuclear separation as the lower state, the product of the two integrals will be identically zero unless $j = k = G$ (i.e. Rayleigh scattering), because the upper state will have the same orthogonal vibrational wavefunctions as the lower one. Even if the vibrational frequency is somewhat different it can be shown that the overlap will be poor, especially for lower vibrational levels.[2] Raman scattering, on this model, arises because of the different equilibrium nuclear separation in the excited state. This means that in order to

[1] C-H Ting, *Spectrochim. Acta*, **24A**, 1177, 1968.
[2] This problem is discussed by A. J. Sadlej, *Spectrochim. Acta*, **26A**, 421, 1970.

calculate the overlap of upper and lower vibrational states, it is necessary to calculate the vibrational wavefunctions of one state about the equilibrium positions of the other one. For example, the ground vibrational state $\psi_G^n(z-\delta)$ turns out to be related to the corresponding states $\psi_j^n(z)$, by the relation

$$\psi_G^n(z-\delta) = \psi_G^n(z) + \frac{\delta}{\sqrt{2}}\psi_1^n(z) + \delta^2\left(\frac{\psi_2^n(z)}{2\sqrt{2}} - \frac{\psi_G^n}{4}\right)$$

$$+ \text{terms in } \delta^3 \text{ and above} \quad \text{(A1.4)}$$

and similarly:

$$\psi_1^n(z-\delta) = -\frac{\delta}{\sqrt{2}}\psi_G^n(z) + \psi_1^n(z) - \delta\psi_2^n(z) + \text{terms in } \delta^2 \text{ and above}$$

$$\text{(A1.5)}[3]$$

Thus we may represent vibrational wavefunctions such as ψ_G^n and ψ_1^n (which describe the initial and final states normally observed in room temperature Raman spectroscopy) as in figure A1.1.

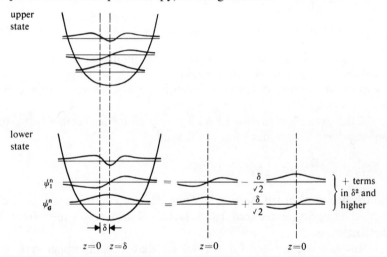

Fig. **A1.1** The rewriting of vibrational wavefunctions about a new point

Once the lower state has been written in terms of wavefunctions centred around the equilibrium position of the upper state (or vice versa), we can immediately write down the vibrational overlap integrals $\int \psi_j^{n*}.\psi_k^n \, d\tau$ for any one state k, by using the orthonormality of the ψ_j^n. Thus our overall transition probability becomes proportional to the square of a sum of readily evaluated overlap integrals.

These sums have been worked out by Ting for a number of Raman possibilities, and in all cases

$$\sum_k \int \psi_j^{n*}.\psi_k^n \, d\tau . \int \psi_k^{n*}.\psi_G^n \, d\tau = 0 \quad \text{(A1.6)}$$

[3] These relations may be readily verified using Taylor's theorem.

unless $j = k = G$. However, the sum

$$\sum_k k \int \psi_1^{n*} . \psi_k^n \, d\tau . \int \psi_k^{n*} . \psi_G^n \, d\tau = \frac{\delta}{\sqrt{2}} \tag{A1.7}$$

In general for a $\Delta n = \pm 1$ transition, the sum in A1.7 is non-zero, but for $\Delta n = \pm 2$ or more, the sum in A1.7 is identically zero. This explains the $\Delta n = \pm 1$ selection rule in Raman spectroscopy. For the sum in A1.7 arises because of the denominator term $\omega_{jk} + \omega$. If ω_{j0} is the frequency equivalent of the energy jump from ψ_j to the ground vibrational level of the upper state, and ω_{0k} is the equivalent of the vibrational energy in the upper state, then

$$(\omega_{jk} + \omega)^{-1} = (\omega_{0k} + \omega_{j0} + \omega)^{-1} = \frac{1}{\omega_{j0} + \omega} \left[1 + \frac{\omega_{0k}}{\omega_{j0} + \omega} \right]^{-1}$$

$$\simeq \frac{1}{\omega_{j0} + \omega} \left[1 - \frac{\omega_{0k}}{\omega_{j0} + \omega} \right] \quad \text{(because } \omega_{0k} \ll \omega_{j0} + \omega\text{)}$$

$$= \frac{1}{\omega_{j0} + \omega} - \frac{k\omega_{01}}{(\omega_{j0} + \omega)^2} \quad \begin{array}{l} \text{(because } \omega_{0k} = k\omega_{01} \text{ for the} \\ \text{harmonic oscillator)} \end{array} \tag{A1.8}$$

This explains the presence of k as an integer multiplier in A1.7, and also predicts that the intensity of a Raman line, all else being equal, will be proportional to $(\omega_{01})^2$, the vibrational frequency. There is some experimental evidence for this latter prediction.

If the expansion leading to A1.8 is carried to one further term, then we also have to consider sums of the kind

$$\sum_k k^2 \int \psi_j^{n*} \psi_k^n \, d\tau . \int \psi_k^{n*} \psi_G^n \, d\tau = 0$$

which are non-zero if $\Delta n = \pm 2$. Thus Raman overtones are predicted whose intensity is reduced by a factor of $(\omega_{01}/(\omega_{j0} + \omega))^2$ from the fundamentals.

In the extreme case of $\delta \ll 1$, we see that the only upper state contributing to the $\psi_G^n \to \psi_1^n$ Raman transition intensity is the $k = 1$ state.

Polyatomics

This approach has so far considered only the selection rules governing changes in n. However, in symmetrical molecules some vibrational modes may undergo no electric dipole transitions at all, and some other modes no Raman transitions.

To understand this fully it is necessary to use the methods of group theory, because this provides a language for describing different symmetries.[4] However, it is possible to describe the special case of centro-

---

[4] See, for example, F. A. Cotton, *Chemical Applications of Group Theory*, John Wiley (Interscience), New York, 1963.

symmetric molecules without using the full language of group theory. The normal vibrations of centrosymmetric molecules fall into two distinct groups, symmetric (g) and antisymmetric (u). In the symmetric vibrations the opposite halves of the molecule move in opposite directions, e.g. v_1 of CO_2, and v_1, v_2, and v_4 of C_2H_2 (figure 9.3). Such modes are inactive in the infrared because they cannot possibly create a temporary molecular dipole moment.

In the antisymmetric vibrations, the opposite halves of the molecule move in the same direction. Examples are v_2, v_3 of CO_2 (the antisymmetric stretching and the bending mode) and v_3, v_4 of C_2H_2. Now antisymmetric vibrations of this type cannot be stimulated by electronic excitation. The linear antisymmetric modes are not stimulated because no possible excited electronic state could give the molecule linear asymmetry. The bent asymmetric modes are not stimulated because, although bent excited states are possible, the initial excitation of a linear molecule must, according to the Born-Oppenheimer principle, produce a vibrationally metastable linear excited state. Thus the electronic excitation cannot in either case give the nuclei any antisymmetric impulse, and so pure antisymmetric states are Raman inactive. Hence the mutual exclusion principle is explained.

This argument would appear also to exclude the v_4 symmetric buckling mode of C_2H_2.[5] However, this mode does appear weakly in the Raman spectrum, probably because of interactions analogous to vibronic coupling in optical spectra (section 10.9). It is generally observed that g-modes which are not totally symmetric, such as v_4 of C_2H_2, give appreciably weaker lines in the Raman than do totally symmetric modes.

---

[5] This is also excluded by the classical analysis, for although the buckling affects α somewhat, $d\alpha/dq$ is zero when the molecule is linear. Also in the normal coordinate analysis α_{xz} and α_{yz} are zero for a linear molecule.

Appendix 2
First-order perturbation theory

Let us assume that we wish to study the effect of a small perturbation whose operator is P on a system whose Hamiltonian operator in the absence of the perturbation is H_0, all of whose (unperturbed) normalized eigenfunctions ψ_n are known. An example might be the effect of an electric field on a hydrogen atom. We thus have

$$H_0\psi_n = \varepsilon_n\psi_n \tag{A2.1}$$

where the ε_n are the energies of the unperturbed states. Let us now write $P = \lambda P_0$. This helps us to study what happens when P is very small by simply varying λ, a pure number. The complete Hamiltonian for the perturbed system is $H_0 + \lambda P_0$, and we wish to discover, at least approximately, its eigenstates $\psi(n)$. If the perturbed energies are $\varepsilon(n)$ then

$$(H_0 + \lambda P_0)\psi(n) = \varepsilon(n)\psi(n) \tag{A2.2}$$

Now when λ becomes vanishingly small, equations A2.2 and A2.1 become identical i.e. $\psi(n) = \psi_n$ and $\varepsilon(n) = \varepsilon_n$. Also as λ increases slightly it is a reasonable assumption that $\psi(n)$ still $\approx \psi_n$ and $\varepsilon(n) \approx \varepsilon_n$, with the small differences between $\psi(n)$ and ψ_n, $\varepsilon(n)$ and ε_n being in some way dependent on λ. Mathematically, this is equivalent to saying that

$$\psi(n) = \psi_n + \lambda\psi_1 + \lambda^2\psi_2 + \text{terms in } \lambda^3 \text{ and above} \tag{A2.3}$$

$$\varepsilon(n) = \varepsilon_n + \lambda\varepsilon_1 + \lambda^2\varepsilon_2 + \text{terms in } \lambda^3 \text{ and above} \tag{A2.4}$$

and also that ψ_1, ε_1, etc. are not very much larger than ψ_n, ε_n. Equations A2.3 and A2.4 are (apart from the final proviso) completely general. They just mean that $\psi(n)$ and $\varepsilon(n)$ are each dependent in some way upon λ. The ψ_1, ψ_2, ε_1, ε_2, etc. are simply correction terms, and are not necessarily eigenfunctions or eigenvalues of anything. They will, of course, be different for each $\psi(n)$. We will now follow the consequences of equations A2.1 to A2.4, knowing that our conclusions will be valid provided that $\psi(n) \approx \psi_n$ and $\varepsilon(n) \approx \varepsilon_n$. Applying A2.3 and A2.4 to A2.2, we obtain:

$$(H_0 + \lambda P_0)(\psi_n + \lambda\psi_1 + \lambda^2\psi_2 + \ldots)$$
$$= (\varepsilon_n + \lambda\varepsilon_1 + \lambda^2\varepsilon_2 + \ldots)(\psi_n + \lambda\psi_1 + \lambda^2\psi_2 + \ldots)$$

that is

$$H_0\psi_n + \lambda H_0\psi_1 + \lambda^2 H_0\psi_2 + \lambda P_0\psi_n + \lambda^2 P_0\psi_1 + \lambda^3 P_0\psi_2 + \ldots$$
$$= \varepsilon_n\psi_n + \lambda\varepsilon_n\psi_1 + \lambda^2\varepsilon_n\psi_2 + \lambda\varepsilon_1\psi_n + \lambda^2\varepsilon_1\psi_n + \lambda^3\varepsilon_1\psi_2 + \lambda^2\varepsilon_2\psi_n + \ldots$$

This equation is still quite general. But we may simplify it by making the approximation that $\lambda \ll 1$, so that terms in λ^2, λ^3 etc. may be set equal to zero in the presence of terms in λ. This approximation is completely valid for small λ, and for larger λ gives us at least the 'first-order' corrections to $\psi(n)$ and $\varepsilon(n)$. We may also note that the first terms on each side of the equation are made equal by equation A2.1. Therefore, putting $\lambda P_0 = P$,

$$\lambda H_0 \psi_1 + P \psi_n = \lambda \varepsilon_n \psi_1 + \lambda \varepsilon_1 \psi_n \tag{A2.5}$$

We now multiply both sides of equation A2.5 by ψ_n^* and integrate over all space; therefore

$$\lambda \int \psi_n^* H_0 \psi_1 \, d\tau + \int \psi_n^* P \psi_n \, d\tau = \lambda \varepsilon_n \int \psi_n^* \psi_1 \, d\tau + \lambda \varepsilon_1 \int \psi_n^* \psi_n \, d\tau$$

But

$$\lambda \int \psi_n^* H_0 \psi_1 \, d\tau = \lambda \int \psi_1^* H_0 \psi_n \, d\tau = \lambda \int \psi_1^* \varepsilon_n \psi_n \, d\tau = \lambda \varepsilon_n \int \psi_n^* \psi_1 \, d\tau$$

and also

$$\int \psi_n^* \psi_n = 1$$

for the ψ_n are normalized. Therefore

$$\lambda \varepsilon_1 = \int \psi_n^* P \psi_n \, d\tau \tag{A2.6}$$

This is the same as equation 10.1. It gives us the first-order correction to the energy ε_k due to the effect of the perturbation P on the state ψ_k.

The calculation of $\lambda \psi_1$, the first-order correction to the wavefunction, also springs from equation A2.5. But first we must extend equation A2.3, by remembering that $\psi(n)$ may also be written, in general, as a series $_j\Sigma a_j \psi_j$ in terms of ψ_j, the complete set of eigenfunctions of H_0. (The ψ_j are the same as the ψ_n, but the different label enables us to concentrate on one state $\psi(n)$ while talking about the contributions to that particular state of all the unperturbed functions ψ_j. Their eigenvalues are ε_j, and the a_j are constants.) We may thus expand equation A2.3 as follows:

$$\psi(n) = \psi_n + \lambda \psi_1 + \lambda^2 \psi_2 + \ldots = \sum_j a_j \psi_j$$

Therefore, provided that terms in λ^2, etc. may be ignored, and provided that $\psi(n) \approx \psi_n$, so that $a_n \approx 1$, we obtain

$$\lambda \psi_1 = \sum_{j \neq n} a_j \psi_j \tag{A2.7}$$

Our sum is now over all states ψ_j except for ψ_n.

Applying A2.7 to A2.5, we obtain

$$H_0 \sum_{j \neq n} a_j \psi_j + P \psi_n = \varepsilon_n \sum_{j \neq n} a_j \psi_j + \lambda \varepsilon_1 \psi_n$$

Now

$$H_0 \sum_{j \neq n} a_j \psi_j = \sum_{j \neq n} \varepsilon_j a_j \psi_j$$

and therefore:

$$P \psi_n = \sum_{j \neq n} (\varepsilon_n - \varepsilon_j) a_j \psi_j + \lambda \varepsilon_1 \psi_n$$

Multiplying all three terms by ψ_k^*, where ψ_k is one particular ψ_j other than ψ_n, and integrating over all space, we obtain:

$$\int \psi_k^* P \psi_n \, d\tau = \sum_{j \neq n} (\varepsilon_n - \varepsilon_j) a_j \int \psi_k^* \psi_j \, d\tau + \lambda \varepsilon_1 \int \psi_k^* \psi_n \, d\tau$$

But ψ_k is orthogonal to all the ψ_j's except itself, and is not ψ_n. Therefore $\int \psi_k^* \psi_n \, d\tau = 0$ and $\int \psi_k^* \psi_j \, d\tau = 1$ if $j = k$ and $= 0$ if $j \neq k$. Hence:

$$a_k (\varepsilon_n - \varepsilon_k) = \int \psi_k^* P \psi_n \, d\tau \quad (k \neq n)$$

We have thus obtained a general formula for a_j provided that $j \neq n$, and hence, using equation A2.7, we see that

$$\lambda \psi_1 = \sum_{k \neq n} \frac{\displaystyle\int \psi_k^* P \psi_n \, d\tau}{\varepsilon_n - \varepsilon_k} \psi_k \tag{A2.8}$$

from which equation 10.2 may be readily derived.

Equation A2.8 predicts that $\lambda \psi_1$ will be small provided that $\varepsilon_n - \varepsilon_k \gg \int \psi_k^* P \psi_n \, d\tau$. If this is not true, then our initial assumption that $\psi(n) \approx \psi_n$ must be invalid, and our theory useless. However, a suitable choice of ψ_k's is usually possible to ensure that $\lambda \psi_1$ is small and hence that equations 10.1 and 10.2 are valid, for all $\psi(n)$, provided that $P \ll H_0$.

Answers to problems

1.1 $R_\infty = 1.097 \times 10^5 \text{ cm}^{-1}$

1.6 $\Delta x \approx 0.05 \text{ nm}$ (using the precise form of the uncertainty principle)

1.7 30 cm^{-1} or 10^{12} Hz

1.8 12 volts

1.9 $\varepsilon = n^2 h^2 / 8\pi^2 m$

1.10 0.08 m s^{-1}

1.13 21.8% (404.0 cm^{-1}) and 0.00% $(102\,406.5 \text{ cm}^{-1})$

3.3 at $90°$ to the sun's direction

3.4 3 m—about four times the length of each arm of a typical VHF aerial

3.6 horizontally

3.7 the planes of the CO_3^{2-} ions must be parallel, on average

4.4 $\alpha = \sum\limits_k 2\varepsilon_k x_{0k}^2 / (\varepsilon_k - \varepsilon_0)^2$

4.7 No

5.1 **a** $13.4 \text{ eV} = h \times (3.24 \times 10^{15} \text{ Hz})$
$= (2.74 \times 10^4 \text{ cm}^{-1}) \times h/c$
b $60 \text{ MHz} = c/(5.0 \text{ m})$
c kT at $25 °C = (207.25 \text{ cm}^{-1}) \times h/c$
d $0.1 \text{ nm} = c/(3 \times 10^{18} \text{ Hz})$
$= c/h \times (1.2 \times 10^4 \text{ eV})$

5.2 25 pm

5.3 **a** u.v., Raman, i.r., n.m.r.
b u.v., electron resonance
c u.v., Raman, electron resonance, n.m.r.
d u.v., electron resonance, n.m.r.

5.7 $1:44.5$ (upper limit); $1:1.37$ (lower limit)

5.9 $4\text{K}: 10^8 \text{ Hz}: 0.999; 10^{10} \text{ Hz}: 0.887;$
$10^{12} \text{ Hz}: 6.16 \times 10^{-6} \left(\dfrac{n_u}{n_1} \right)$
$300\text{K}: 10^8 \text{ Hz}: 1.000; 10^{10} \text{ Hz}: 0.998;$
$10^{12} \text{ Hz}: 0.852 \times 10^{-6} \left(\dfrac{n_u}{n_1} \right)$

5.10 $5 \times 10^2 \text{ T}$ $(5 \times 10^6 \text{ gauss})$. Such fields are not at present stably attainable, and spin pumping techniques have

to be used to lower the effective spin temperature still further.

5.11 $\dfrac{kT}{h} \ln\left(\dfrac{u}{u+1} \right) (\text{Hz})$ or
$\dfrac{kT}{100\,hc} \ln\left(\dfrac{u}{u+1} \right) (\text{cm}^{-1})$

6.3 Possibly: field inhomogeneity; scan too fast (ringing); ditto (time constant distortion); poor setting of phase detector or of orthogonal pick-up coil; two resonances observed; machine instability.

6.4 $9.20 \times 10^{10} \text{ Hz}$, 0.0045 nm

7.1 11.19 MHz, $7.04 \times 10^7 \text{ T}^{-1} \text{s}^{-1}$

7.2 2.135

7.9 nearly planar (some vibronic distortion)

7.10 four orientation-dependent resonances. At high temperature there will be a dynamic Jahn–Teller effect.

7.11 $CH_3.\dot{C}H.OH$ with hydroxyl proton exchanging rapidly.

7.12 $VO_4^{3-} \xrightarrow{\text{H}^+} VO_4H^{2-} \rightarrow V_2O_7^{4-} \rightarrow$
$\phantom{VO_4^{3-}} {}^{\text{H}^+}\!VO_3^-$
$\xrightarrow{\text{H}^+} V_2O_7H^{3-} \xrightarrow{\text{H}^+} ?V_3O_9^{3-}$

8.1 **a** $B = 19\,976.04 \text{ MHz}$,
$D = 65–70 \text{ kHz}$
b the masses given in the problem are incorrect: they should be
$^1H = 1.6734$, $^2H = 3.3441$,
$^{12}C = 19.9250$, $^{31}P = 51.4626$, all
$\times 10^{-27} \text{ kg}$. With these masses,
$C{-}H = 0.1067 \text{ nm}$ and
$C{\equiv}P = 0.1543 \text{ nm}$.

8.6 three

8.11 $^2H_2: 1$ odd to 2 even
$(^1H^{13}C)_2: 3$ odd to 5 even
These answers refer to the simple case

of a symmetric ψ_{vib} and ψ_{elec}. If either of these were to change sign upon rotation of the molecule through 180°, then 'odd' and 'even' would interchange.

9.3   **a** $\Delta H = +0.73\,\text{kJ mol}^{-1}$. This would be unaffected by molecular rotation unless $T \geqslant 100\text{K}$, in which case there would be a specific heat contribution from rotation which would be greater for HD because of the absence of spin restrictions on the rotational states.

9.4   linear, NNO

9.7   0.81% increase from $n = 0$ to $n = 1$

10.3   transition moment will be along the bond axis

10.4 ^5S

10.6   **a** and **b** forbidden, allowed. Isotropic transition moment

   c   allowed (t.m. parallel to bond), forbidden

   d   weakly allowed (vibronic coupling), t.m. isotropic for symmetric complex.

10.7   Transition moment normal to bond axis. Probably a $\sigma^* \leftarrow \pi^*$ transition, but some d-orbital participation may be involved.

10.15 $152.5\,\text{kJ mol}^{-1}$

Index

The most important references to major topics are given in bold type. References to figures or problems only are in italics. Some symbols are included in the index, the single reference being to the page on which they are defined or introduced. A few references are given to pages on which the subject is discussed only implicitly, or only under a different name.